Lewis Hamilton

A Silver Legacy at Mercedes

Etienne Psaila

Lewis Hamilton: A Silver Legacy at Mercedes

Copyright © 2024 by Etienne Psaila. All rights reserved.

First Edition: **December 2024**

No part of this publication may be reproduced, distributed, or transmitted in any form or by any means, including photocopying, recording, or other electronic or mechanical methods, without the prior written permission of the publisher, except in the case of brief quotations embodied in critical reviews and certain other non-commercial uses permitted by copyright law.

ISBN: 978-1-923393-29-5

Table of Contents

1. The Decision to Move (Late 2012–Early 2013)
2. Settling In: The 2013 Season
3. A New Era Begins: 2014 and the Turbo-Hybrid Revolution
4. Consolidating Dominance: 2015
5. The Intense Duel: 2016
6. A New Partnership: 2017 with Bottas
7. Chasing and Breaking Records: 2018–2019
8. Equalling a Legend: The 2020 Season
9. The Narrow Defeat: 2021's Title Showdown
10. Struggling with Change: The 2022 Regulations Shift
11. Partial Recovery and Ongoing Challenges: The 2023 Season
12. Last Season together: The 2024 Season
13. Technical Innovations and Car Development
14. Team Management and Strategy
15. Key Rivalries and Partnerships
16. Records and Milestones
17. Cultural Impact and Media Response
18. Safety and Sporting Regulation Changes
19. Hamilton's Driving Evolution and Technical Feedback
20. Mentorship and Influence Within the Team
21. Hamilton's Evolving Brand and Off-Track Engagement
22. Circuit Mastery and Signature Performances
23. Adapting to Psychological and Competitive Pressures
24. A Farewell and a New Horizon
25. A Gallery of Iconic Moments

Chapter 1: The Decision to Move (Late 2012–Early 2013)

Lewis Hamilton's departure from McLaren at the end of 2012 marked one of the most significant shifts of his early Formula One career. Until that point, Hamilton had been inextricably linked with McLaren. They had scouted him as a gifted young karter, nurtured him through the junior formulas, and given him his F1 debut in 2007. In just his second season at the pinnacle of motorsport, he claimed the 2008 World Drivers' Championship under their banner. By the close of the 2012 campaign, he had spent six consecutive seasons in McLaren's race seat—enough time to be regarded as a fixture in the team's history.

The announcement of Hamilton's move to Mercedes AMG Petronas came on September 28, 2012. He would be joining the German manufacturer's works team starting with the 2013 season, replacing the seven-time champion Michael Schumacher, who was retiring. To the public, this decision might have seemed bold, even risky. McLaren, after all, was a proven front-runner, a team with an extensive record of victories and championships. Mercedes, in contrast, had returned as a works outfit only in 2010 and, while showing occasional flashes of promise, had secured just a single Grand Prix win in three years. Many observers questioned whether the move would benefit Hamilton's career in the short term.

Behind the scenes, the reasoning was more nuanced. Hamilton was looking toward a future that extended beyond the current competitive order. He recognized that Formula One was on the brink of a regulatory overhaul set to arrive in 2014, one that would usher in a new era of turbo-hybrid engines. Mercedes, as both a constructor and an engine manufacturer, held the potential to excel in this new technical landscape. Senior figures like Team Principal Ross Brawn and the newly appointed non-executive chairman Niki Lauda actively courted Hamilton, presenting a vision of long-term ambition and the resources needed to realize it. Their message was simple: Mercedes intended to become a championship-winning force.

Financial details also played a role. Hamilton's contract with Mercedes was reportedly lucrative, though exact figures were not publicly confirmed at the time. More importantly, the arrangement offered him a clear leadership position within the team's driver lineup. With Schumacher retiring and Nico Rosberg already in place, Hamilton would step into an environment eager to build around him rather than one defined by McLaren's longstanding internal culture and deep driver development pipeline.

As 2012 drew to a close, Hamilton fulfilled his final obligations with McLaren. The 2012 season concluded with the Brazilian Grand Prix, where he completed his last race for the team that had shaped his early career. McLaren offered public statements of respect and gratitude, acknowledging Hamilton's contributions and the world title they had achieved together. For Hamilton, the farewell was bittersweet: he left with 21 Grand Prix victories and a drivers' crown in McLaren colors, but also a sense that his next chapter might hold untapped potential.

When the testing period for 2013 began, Hamilton donned the silver-and-teal overalls of Mercedes for the first time. He climbed into the cockpit of the W04, the team's new challenger, and set about adapting to his fresh surroundings. Early tests were encouraging rather than spectacular, indicating that the upcoming season would be one of transition. Still, there was a palpable energy within the Mercedes garage and a collective sense that the team's substantial investment in personnel and technology would soon bear fruit.

Hamilton's move to Mercedes was, from any factual standpoint, a calculated decision with long-term goals in mind. He arrived not with the expectation of immediate dominance, but with the conviction that Mercedes had the infrastructure, talent, and foresight to climb to the top of Formula One. The late months of 2012 and the dawn of 2013 were marked by the hum of preparation and possibility—a new driver in a new team, laying the groundwork for what many hoped would be a transformative era to come.

Chapter 2: Settling In: The 2013 Season

When the Formula One circus convened at Albert Park in Melbourne for the opening race of 2013, Lewis Hamilton stepped onto the grid wearing Mercedes colors for the first time. Although the previous few months had been filled with speculation, contract discussions, and media scrutiny, the moment of truth arrived under the Australian sun on March 17. Hamilton, paired with Nico Rosberg, now represented the Mercedes AMG Petronas team—an outfit that had shown potential but not yet proven itself as a consistent front-runner since its 2010 return as a works squad.

In testing, the Mercedes W04 had looked steady rather than explosive, its sleek silver lines concealing a car that was still learning to spread its wings. Guided by Team Principal Ross Brawn and powered by the proven Mercedes V8 engine, the team's technical personnel, including figures like Aldo Costa and Geoff Willis, aimed to refine the aerodynamic package and achieve a balance that would allow Hamilton to exploit his well-documented late-braking, high-corner-entry-speed style.

The first few races provided an early measure of progress. Hamilton's debut with Mercedes in Australia yielded a fifth-place finish, a respectable result that confirmed the team could hold its own in the congested midfield battle. One race later in Malaysia, Hamilton took his first podium for Mercedes, crossing the line in third place. While not yet at the level of the Red Bulls or Ferraris, the team's steady improvements were becoming visible. The handling issues that had plagued them in 2012 seemed less acute, and Hamilton was quick to acknowledge the team's efforts in extracting more performance from the car's chassis and tires.

The early European races strengthened this upward trajectory. At the Chinese Grand Prix in April, Hamilton secured pole position—his first pole in Mercedes colors—and followed up with another podium finish, demonstrating that not only could he qualify at the sharp end, but the team could also manage race pace with reasonable efficiency. The result was more than just another trophy for the cabinet; it was a clear signal that Mercedes was evolving into a serious competitor, inching closer to the leaders rather than slipping behind.

Throughout the summer months, incremental gains kept coming. Engineers focused on understanding Pirelli's sensitive tires and refining the W04's aerodynamics. Hamilton's driving style, honed at McLaren, adapted to the new environment. He worked closely with Rosberg and their race engineers to find the ideal setup window, piecing together data from practice sessions, simulations, and wind tunnel results. Although there were still off-weekends—Formula One's intensity ensures no straight-line progression—there was a palpable sense within the team that the path forward was clear and stable.

A highlight of Hamilton's first season with Mercedes came in late July at the Hungarian Grand Prix. Circuit Hungaroring, a tight, twisty track often likened to a karting circuit for its lack of long straights, seemed tailored to Hamilton's precise driving style. Starting from pole position, he controlled the race pace and, managing both tire wear and strategy, took the checkered flag to claim his first victory with Mercedes. This triumph was no fluke; it was a culmination of the season's progressive steps, the result of hundreds of engineering refinements, strategy meetings, and the driver's relentless push for perfection.

Beyond the podium ceremonies and the champagne sprays, 2013 also laid the organizational groundwork for future successes. Internally, Mercedes was coalescing into a more cohesive unit. The technical team had begun to find its rhythm, communicating effectively and responding swiftly to new challenges. Toto Wolff, who joined the team's management structure early in the year, brought a business-minded approach that meshed with the existing technical and racing expertise. Niki Lauda, as non-executive chairman, continued to encourage an environment of accountability and ambition.

By the end of the season, Hamilton finished fourth in the Drivers' Championship, amassing 189 points, while Mercedes rose to second in the Constructors' standings—its highest finish since returning to the sport as a full factory effort. Considering the team had only managed a sole victory in three years prior, this leap into the upper echelons of the field was both encouraging and instructive. It underscored that the driver's move from McLaren, considered risky by some, was indeed setting the stage for something larger and more defining.

The foundation had been laid. The 2013 season didn't just provide immediate results in the form of podiums and a breakthrough win; it served as an invaluable period of learning and refinement. Within the silver-clad garages, there was a collective understanding that while they had come far, the true test lay ahead. Hamilton's assimilation into Mercedes had proven fruitful, and as the final checkered flag of 2013 fell, it was clear that both driver and team were aligning their strengths. This alignment would prove critical as Formula One's era was about to change dramatically, and in that change, Mercedes would find an opportunity to define its legacy.

Chapter 3: A New Era Begins: 2014 and the Turbo-Hybrid Revolution

The start of the 2014 Formula One season heralded one of the most significant technical revolutions in the sport's modern history. Gone were the naturally aspirated V8 engines that had been the norm for nearly a decade. In their place came 1.6-liter V6 turbo-hybrid power units, intricate systems blending internal combustion engines with advanced energy recovery components. Many experts anticipated that these sweeping regulation changes would reorder the competitive landscape, and Mercedes, as both a constructor and an engine supplier, had prepared meticulously for this moment.

From the very first pre-season tests, the Mercedes AMG Petronas team's meticulous engineering efforts became evident. The new car, the F1 W05 Hybrid, showcased not only a powerful and reliable engine but also a chassis and aerodynamic package that worked in near-perfect harmony. The team's engineering group, led by figures like Technical Director Paddy Lowe and backed by the resources of Mercedes High Performance Powertrains in Brixworth, delivered a unit that blended outright power with energy efficiency. While rivals such as Renault and Ferrari wrestled with reliability issues and drivability problems, Mercedes' powertrain ran with remarkable consistency from day one.

Lewis Hamilton, now in his second season with Mercedes, approached 2014 as a pivotal juncture in his career. He had already proven he could adapt, but this was a fresh challenge: brand-new technology, new handling characteristics, and a recalibrated playing field. Early on, it became clear that the W05 offered not just a competitive car, but a dominant one. Both Hamilton and teammate Nico Rosberg found themselves at the sharp end of the grid regularly, locking out front rows in qualifying and executing lights-to-flag performances on race day.

The season opener in Australia established the tone. Although Hamilton retired from that race due to a technical issue, Rosberg cruised to victory, confirming that the Silver Arrows had a potent weapon. In the subsequent rounds, Hamilton struck back. Victories in Malaysia, Bahrain, China, and Spain came in quick succession. The Bahrain Grand Prix, in particular, showcased the team's supremacy; Hamilton and Rosberg fought fiercely for the lead, yet nobody else could come close. The rivalry between the teammates made for gripping viewing, but it also underscored just how far ahead Mercedes was of the chasing pack.

While Hamilton's natural speed and racecraft were as strong as ever, what made 2014 truly special was the synergy between driver and machine. The turbo-hybrid power unit demanded careful energy management. Drivers had to balance the internal combustion engine's output with the electrical energy harvested under braking and deployed on the straights. Hamilton proved adept at this balancing act, using Mercedes' sophisticated steering wheel controls and on-board systems to maintain the optimal flow of power. On high-speed circuits like Monza, the W05's horsepower advantage was evident; on twistier layouts, the car's aerodynamic efficiency and torque delivery gave it unparalleled drive out of slower corners.

Throughout the year, the Hamilton-Rosberg battle became the center-stage storyline. Rosberg's victories in Australia and Monaco, combined with a series of pole positions, ensured that Hamilton was never able to relax. Yet Hamilton's response was unwavering. He captured eleven wins in total, demonstrating an edge in wheel-to-wheel combat. When controversy arose—like the collision between the two teammates in Belgium—Hamilton recovered with a calm determination that illustrated his growing maturity. He refused to be derailed, stringing together a late-season run of victories that placed him firmly in the championship lead.

The Drivers' Championship came down to the final race in Abu Dhabi, where a new double-points rule meant Rosberg still had a mathematical chance. Under the floodlights of the Yas Marina Circuit, Hamilton made a perfect start from second on the grid, took the lead into the first corner, and never looked back. Rosberg, hobbled by a technical problem, faded down the order. Hamilton crossed the finish line as the 2014 World Drivers' Champion—his second career title, and importantly, his first with Mercedes.

There were other accolades, too. Mercedes clinched the Constructors' Championship well before the end of the season, amassing a record-breaking points tally. The team achieved 16 wins out of 19 races and racked up numerous front-row lockouts. Engineers in Brackley and Brixworth had their work vindicated, establishing a new era of dominance. For Hamilton, the season reinforced that his choice to join Mercedes in 2013 had been a masterstroke. He had placed himself at the center of a team engineered for success in a time of unparalleled change.

By the end of 2014, it was clear that a new era in Formula One had indeed begun, and Mercedes was its benchmark. The phrase "turbo-hybrid revolution" had a real-world embodiment in the Silver Arrows' relentless performance, and at the helm of it all stood Lewis Hamilton—champion once again, and this time as part of a team poised to rewrite the record books.

Chapter 4: Consolidating Dominance: 2015

The year 2015 opened with an air of expectancy at the Mercedes AMG Petronas headquarters in Brackley. The team was riding high from its triumphant 2014 campaign, which had seen Lewis Hamilton secure the Drivers' Championship and Mercedes claim the Constructors' title with unprecedented authority. Yet, Formula One rarely stands still, and the team's key figures—Toto Wolff, Paddy Lowe, and Niki Lauda among them—were determined not to let complacency seep into their ranks. The question facing Mercedes was whether they could maintain their remarkable standards for a second straight season in the turbo-hybrid era.

The new car, the F1 W06 Hybrid, debuted in winter testing with an assurance that bordered on the serene. While rival teams hoped for regulatory tweaks or developmental leaps to challenge the Silver Arrows, early indications showed Mercedes still held a clear performance advantage. The engine unit remained both powerful and fuel-efficient, while aerodynamic refinements and careful packaging improvements underlined the team's relentless pursuit of marginal gains. Internally, there was a well-structured process of continuous feedback and iteration—engineers and mechanics collaborated seamlessly, benefiting from stable leadership and a corporate culture that rewarded open communication and incremental improvement.

On the driver front, continuity was key. Lewis Hamilton and Nico Rosberg returned for their third season together at Mercedes. Both had learned lessons from the intense 2014 duel. Rosberg had gained insight into the mettle required to topple a champion-level teammate, while Hamilton understood the importance of converting every advantage into tangible results. Their rivalry still simmered beneath the surface, occasionally flaring into tense on-track moments, but management made it clear that internal friction would not be allowed to destabilize the collective mission.

From the outset of the racing calendar, Mercedes reasserted its supremacy. At the season-opening Australian Grand Prix, Hamilton and Rosberg delivered a crushing one-two finish, several steps ahead of their nearest challengers. Ferrari, with Sebastian Vettel now on board, did provide a sterner test than in 2014, scoring a surprise win in Malaysia when Mercedes struggled with tire temperatures in the tropical heat. This lone misstep, however, served to sharpen Mercedes' focus. The team responded by refining race strategies, improving tire management techniques, and ensuring that any vulnerabilities exposed by their rivals were addressed promptly.

Hamilton settled into a groove of confident, controlled performances. He claimed ten Grand Prix victories over the course of the season, from familiar haunts like Silverstone—where he triumphed once again in front of an enthusiastic British crowd—to new additions such as the returning Mexican Grand Prix at a revitalized Autódromo Hermanos Rodríguez. His driving reflected a balance of outright speed and strategic acumen, making use of the W06's strengths and working closely with his engineers to extract every drop of performance from both chassis and power unit.

By the time the championship reached Austin, Texas, for the United States Grand Prix in October, Hamilton had an opportunity to clinch his second consecutive Drivers' Championship with Mercedes. In a race affected by rain and unpredictable conditions, he battled Rosberg and Vettel, eventually taking the checkered flag and the title in one fell swoop. That afternoon at the Circuit of the Americas was both a confirmation and a celebration—confirmation that Hamilton had elevated himself to a new level of competitiveness, and celebration of the unity that had formed within the team.

Meanwhile, Rosberg's strong performances, including several late-season wins, helped the team secure its second straight Constructors' Championship well before the final races. Unlike the previous year's inter-team tensions, 2015 concluded with a somewhat more harmonious balance. While the rivalry between Hamilton and Rosberg did not disappear, the collective endeavor to maintain dominance encouraged a stronger professional respect. It was understood that both drivers, in pushing each other to excel, also propelled Mercedes to greater heights.

Off the track, Mercedes continued to invest in its personnel and infrastructure. The human capital of the team—engineers, aerodynamicists, strategists, mechanics—remained stable, allowing knowledge and expertise to deepen year over year. Sophisticated simulator work and data analysis refined every aspect of race preparation. Communication protocols ensured that lessons learned on Sunday afternoons fed directly into the next iteration of design and strategy. This was how a winning culture became embedded: not just by savoring victories, but by systematically understanding and replicating the conditions that produced them.

By season's end, the numbers painted a clear picture: Hamilton had achieved back-to-back world titles,

Mercedes had secured consecutive Constructors' Championships, and the team's points tallies dwarfed those of even its nearest challengers. This was no mere follow-up to 2014's breakthrough dominance; it was proof that Mercedes had laid down long-term markers. They had become the benchmark against which other teams measured themselves, a modern powerhouse that thrived in F1's most technologically complex era.

As the paddock began its winter migration, there was little doubt that Mercedes had consolidated its dominance in 2015. The team had transformed from an ambitious challenger into a well-oiled, title-winning machine. In doing so, it had fortified a culture that prized innovation, discipline, and adaptability—values that would carry it forward as it continued to leave a defining mark on Formula One's new era.

Chapter 5: The Intense Duel: 2016

By the time the 2016 season dawned, Mercedes had already established itself as the dominant force of the turbo-hybrid era. Consecutive championships in 2014 and 2015 set the stage for another title fight. The intrigue, however, did not center on whether Mercedes would reign supreme over other teams—most observers regarded that as a foregone conclusion—but rather on the escalating rivalry between Lewis Hamilton and Nico Rosberg, the two drivers tasked with extracting the utmost from the team's formidable machinery.

The Mercedes F1 W07 Hybrid took to the track looking like a refined successor to its predecessors. Engineers had honed the aerodynamic package, further integrated the power unit's advanced energy recovery systems, and dialed in suspension settings that broadened the car's operating window. In testing, the W07 ran efficiently, its reliability and performance seemingly setting it apart from any other chassis-engine combination. Internally, the technical staff was confident that they had produced yet another championship-caliber machine, and the early rounds bore that out.

It was within this context that the rivalry between Hamilton and Rosberg flared to new heights. Rosberg started the season in commanding form. Capitalizing on Hamilton's occasional poor starts and misfortunes, Rosberg rattled off a sequence of wins—four victories in the opening four races—seizing an early, substantial lead in the Drivers' Championship. Hamilton, frustrated but far from defeated, knew he had the speed to claw back the deficit. The question was whether he could do so cleanly or whether the battle lines would fracture the team's carefully maintained unity.

The tension manifested dramatically at the Spanish Grand Prix. On the opening lap, with both Silver Arrows aiming to assert dominance, Rosberg and Hamilton collided, sending both cars careening into the gravel trap and out of the race. The incident was a flashpoint that brought the team's internal rivalry into sharp focus. While Mercedes management, led by Toto Wolff and Niki Lauda, maintained that both drivers had free rein to compete, this freedom came with an expectation of respect and a shared responsibility not to sabotage the team's race prospects. The Spain collision underscored the fine line both drivers walked.

Despite this setback, Mercedes continued to tower above the competition. Ferrari and Red Bull occasionally threatened to steal a win, as Red Bull did in Spain after the Mercedes double-DNF, but these moments were exceptions rather than the norm. The W07's blend of horsepower, efficiency, and aerodynamic sophistication ensured that, over the course of a season, no rival team could match its pace consistently. The technical unit back at Brackley and Brixworth provided steady updates—improving engine maps, refining the floor, and adjusting brake-by-wire systems—to stay ahead of the incremental gains made by their pursuers.

For Hamilton and Rosberg, the year developed into a strategic tug-of-war. Mid-season, Hamilton mounted a robust comeback, stringing together a series of victories that eroded Rosberg's points advantage. By the summer break, the title fight looked finely balanced. Every qualifying session, every race start, and every pit stop carried a heightened tension. Some weekends, the difference between triumph and disappointment came down to the smallest detail: a slower getaway from the grid, a marginally compromised race strategy, or the ability to manage tires at critical moments.

Mechanical issues also played a role. Hamilton suffered reliability setbacks that cost him points, most notably in Malaysia, where an engine failure forced him out while leading. Rosberg, meanwhile, faced his own pressures—knowing that to secure the championship, he had to absorb Hamilton's relentless pursuit without cracking. It became clear that the rivalry was not simply a matter of driving skill; it was about mental resilience and the capacity to maintain composure under intensifying scrutiny.

The showdown arrived in Abu Dhabi, the season's final race. Hamilton entered the weekend trailing Rosberg in the Drivers' standings. Even if Hamilton won, Rosberg needed only a podium finish to secure the title. After Hamilton took the lead in the race, he employed a controversial tactic, slowing his pace in an attempt to bunch Rosberg into the clutches of Sebastian Vettel and Max Verstappen behind. Rosberg, aware of the dangers, navigated the situation deftly. He finished second, enough to clinch his first and only World Drivers' Championship. Hamilton's late gambit fell short, and although the team had long since wrapped up the Constructors' crown for a third straight year, the final result of the Drivers' contest remained a source of intense emotion within the garage.

For Mercedes, 2016 was a case of maintaining supremacy while managing the delicate human element at its heart. The W07 delivered on its promise, capturing

19 wins from 21 races and another Constructors' title by a comfortable margin. It established an enviable record that secured Mercedes' place as one of the most formidable teams in the sport's history.

Yet the year will forever be remembered for the fierce internal struggle between two evenly matched teammates. The intense duel saw Rosberg emerge as champion, only for him to retire days later, leaving Hamilton as the remaining cornerstone of the team's future ambitions. For Mercedes, the 2016 season reinforced that managing intra-team rivalry was as critical to success as technical prowess. The team's architects, both on the pit wall and in the design office, emerged with fresh insights into balancing performance with human dynamics. In doing so, they prepared to enter the next chapter of their remarkable journey at the pinnacle of Formula One.

Chapter 6: A New Partnership: 2017 with Bottas

With the dust barely settled on the dramatic events of 2016, Mercedes found itself facing an unforeseen and urgent challenge: replacing the newly crowned World Champion Nico Rosberg. Just days after clinching his first and only Drivers' title, Rosberg stunned the motorsport world by announcing his retirement from Formula One. This sudden departure forced Mercedes' leadership—Toto Wolff and Niki Lauda chief among them—into swift action. Their chosen successor was Valtteri Bottas, a talented Finnish driver extracted from his Williams contract to join Lewis Hamilton at the reigning champions. For Mercedes, 2017 would be about forging a new partnership and facing a new era of technical regulations and competition.

The rules for 2017 had been extensively revised, resulting in wider tires, broader front wings, and more aggressive aerodynamic profiles. Cars looked more imposing on track, and engineers had to recalibrate their understanding of downforce, drag, and mechanical grip. The Mercedes F1 W08 EQ Power+ emerged from the design office as a sleek but complex machine. Its longer, wider, and more aerodynamically efficient chassis promised high peak performance, yet early testing indicated it could be sensitive to setup changes and tire management—factors that would demand versatility and cooperation from both drivers and engineers.

Entering the season, Hamilton and Bottas presented a contrast to the tension of years past. Bottas arrived with a reputation for clean, consistent driving, earned through steady points-scoring and several podium finishes at Williams. Though he was stepping into a championship-winning team, he exuded a calm, businesslike demeanor that contrasted the sometimes fractious Hamilton-Rosberg dynamic. From the outset, there was mutual respect. Hamilton welcomed Bottas with encouragement, and Bottas, in turn, embraced the enormous responsibility of racing for the best team in the sport. Still, the internal pecking order was untested, and the world waited to see how quickly this new partnership would gel.

On the competition front, 2017 brought a revived challenge from Ferrari. The Italian team, led by Sebastian Vettel, arrived at the season opener in Australia with a car capable of trading blows with Mercedes. For the first time in the hybrid era, another constructor's machine genuinely threatened to derail the Silver Arrows' supremacy. Early on, Vettel seized victories in Australia and Bahrain, while Hamilton and Bottas countered with wins in China and Russia, respectively. Bottas's maiden Grand Prix victory in Sochi was particularly significant—proof that he could handle pressure at the sharp end and deliver when opportunities arose.

This season quickly became a more complex game of strategy, development, and endurance. Unlike the previous years, where Mercedes often sat comfortably ahead, now the team had to innovate and respond rapidly. Engineers in Brackley worked tirelessly to adapt to the car's sometimes narrow operating window, introducing aerodynamic updates and refining suspension components to maximize tire performance. The trackside operations crew, under relentless scrutiny, honed strategies aimed at countering Ferrari's improved race pace. Pit stops, fuel management, and the timing of tire changes became pivotal elements that could sway a Grand Prix's outcome.

Meanwhile, Bottas and Hamilton established a working relationship founded on clear communication and a shared desire to maintain Mercedes' winning culture. Bottas, acutely aware he was in one of the most coveted seats in motorsport, took every podium and front-row start as a chance to reinforce his value. Hamilton pushed relentlessly, motivated not only by the quest for a fourth title but also by the invigorated challenge posed by Vettel. While the rivalry at the front now stretched across team lines rather than within the same garage, the intra-team atmosphere remained constructive. Team briefings and debriefs were marked by a focus on collective improvement rather than personal scores to settle.

The championship narrative ebbed and flowed. Vettel's early-season lead put pressure on Mercedes to respond. Over the summer races, the Silver Arrows worked methodically to unlock more consistent performance from the W08. After the mid-season break, Hamilton found a new gear—winning crucial races in Belgium, Italy, and Singapore. Bottas, too, contributed valuable points and even took victory in Abu Dhabi at the season's close. Despite Ferrari's best efforts, small misfortunes, mechanical gremlins, and strategic missteps cost them ground in the title fight. By the time Formula One arrived in Mexico late in the year, Hamilton sealed the Drivers' Championship, his fourth overall and third with Mercedes. The team secured its fourth consecutive Constructors' crown soon after, underscoring its enduring excellence.

For Mercedes, 2017 represented a vital test of adaptability. They had lost one champion and gained another talented driver, faced more stringent competition, and dealt with the complexity of new technical rules. Through it all, they emerged still at the top—just not by the unassailable margin of years past. The Bottas-Hamilton partnership proved functional and respectful, allowing the team to focus on countering external threats rather than managing internal friction. In many respects, the season served as a reminder that success in Formula One depends on balancing countless variables—human, technical, and strategic—and that maintaining championship form requires continuous evolution.

As the paddock packed up for the winter, Mercedes could reflect on a season defined by adaptation and resilience. The 2017 campaign had tested the team's unity, engineering prowess, and ability to nurture a new driver pairing under heightened pressure. Once again, they had passed that test, laying the groundwork for the challenges still to come.

Chapter 7: Chasing and Breaking Records: 2018–2019

As Formula One hurtled into the 2018 season, the Mercedes AMG Petronas team was fresh from having secured four straight Constructors' and Drivers' titles. Yet, the prospect of easing back was never on the agenda. The competition had tightened significantly, with Ferrari and Sebastian Vettel proving themselves increasingly formidable rivals. For Mercedes, maintaining its winning streak demanded not only technical excellence but also the tenacity to withstand fierce pressure both on and off the racetrack.

In 2018, the championship battle once again centered on Lewis Hamilton versus Vettel. Early on, Ferrari showcased strong one-lap pace and improved race-day execution. Vettel drew first blood in Australia and Bahrain, sparking speculation that the long-standing Mercedes dominance might waver. The W09—the latest Silver Arrow—was a refined evolution of its predecessor, but it occasionally struggled to find an optimal balance in varying conditions. Some circuits favored Ferrari's aerodynamic efficiency or tire management, forcing Mercedes to address subtle weaknesses and introduce development upgrades at a steady pace.

Hamilton, meanwhile, tapped into a wealth of experience. He won crucial races at key junctures: a rainy qualifying session in Hungary, a remarkable overtaking display at Monza where Ferrari was favored, and a measured victory in Singapore. Each success was the product of a collaborative effort—race engineers calculating pit stop strategies, aerodynamicists improving downforce distribution, and the garage crew executing flawless stops under the intensity of a close title fight. As summer turned to autumn, Mercedes found a consistency that Ferrari struggled to match. Costly errors, mechanical issues, and strategic miscalculations chipped away at Ferrari's early momentum.

By the time the paddock arrived in Mexico City for the late-season showdown, Hamilton stood on the brink of a fifth Drivers' Championship. Securing the title there meant joining an exclusive club of multiple-time champions that included Fangio, Prost, Schumacher, and Vettel himself. With controlled aggression, Hamilton sealed the deal, and the team wrapped up its fifth consecutive Constructors' crown. The feat reaffirmed Mercedes' status as the benchmark in Formula One's hybrid era—an achievement underpinned by relentless development, data-driven decision-making, and cohesive teamwork.

If 2018 was a tense duel that tested Mercedes' mettle, 2019 brought more records into view. The regulations remained relatively stable, and the W10 emerged as an even more integrated and efficient package. Designers at Brackley had refined aerodynamic concepts and improved mechanical grip, while the power unit from Brixworth continued to push the boundaries of hybrid efficiency and raw horsepower. Hamilton, now a five-time champion, and Valtteri Bottas, more confident in his second full season with the team, found themselves commanding the field with a series of early-season one-two finishes that astonished the paddock.

Ferrari did pose challenges—especially at power-sensitive circuits—and Red Bull, armed with a new Honda engine supply, occasionally emerged as a threat. Still, Mercedes sustained a level of consistency that left rivals searching for answers. The first half of 2019 was marked by a remarkable series of Mercedes victories, each contributing to a swelling points tally and further tightening the team's grip on the championships. Bottas scored valuable wins and podiums, reinforcing his credentials as a top-tier driver and bolstering the team's assault on both titles.

Hamilton excelled at measuring the season's rhythm. He collected wins under pressure, managed tires when required, and capitalized on opportunities where others faltered. As the calendar's end approached, Hamilton traveled to the United States Grand Prix with the chance to secure his sixth Drivers' Championship. By finishing second in Austin, Texas, he locked in that historic achievement. With that result, Hamilton leapt ahead of Juan Manuel Fangio's total and placed himself second only to Michael Schumacher's then-record of seven titles. Days later, Mercedes claimed its sixth consecutive Constructors' Championship, extending its unprecedented run and confirming its place among the greatest dynasties in motorsport history.

The numbers told a story of near-absolute control: by 2019's conclusion, Mercedes had earned more than 100 Grand Prix victories since re-entering Formula One as a works team in 2010 and had assembled an unparalleled streak of successive titles. Each increment—every podium, pole position, and championship—established new benchmarks against which future teams and drivers would be measured. Beyond the statistics, however, lay a core philosophy that had carried Mercedes through every challenge. The team had nurtured a culture of accountability, constant improvement, and technical ingenuity. It recognized that every pit crew member, strategist, aerodynamicist, and engine designer contributed to its remarkable journey.

As 2019 ended, Mercedes had chased and broken records at a relentless pace. Their successes over Vettel and Ferrari were no longer narrow escapes, but rather affirmations that the team had perfected the art of performance under sustained pressure. Surpassing milestones had become as much a part of Mercedes' DNA as the three-pointed star on its badge. The stage was set for what would come next—fresh challenges, new innovations, and the unending pursuit of excellence that defined the Silver Arrows' place in motorsport's grand narrative.

Chapter 8: Equalling a Legend: The 2020 Season

The 2020 Formula One season began under circumstances unlike any before. As the COVID-19 pandemic gripped the globe, race calendars were upended, and the usual rhythms of testing, team briefings, and travel were replaced by uncertainty and emergency planning. Formula One's organizers, teams, and the FIA worked together to craft a revised schedule that began in July, four months later than originally intended, and took place largely without spectators. For Mercedes, the challenge was to maintain peak performance in an environment where long breaks, stringent health protocols, and a condensed race calendar demanded remarkable adaptability.

Into this uncertain landscape rolled the Mercedes F1 W11 EQ Performance. Under the direction of Technical Director James Allison and guided by Toto Wolff's leadership, the team had spent the off-season refining an already dominant platform. Notably, the W11 integrated the innovative Dual-Axis Steering (DAS) system, allowing the drivers to adjust the toe angle of the front wheels from the cockpit. While this was not the sole key to dominance, it exemplified the kind of engineering cleverness that often separates leaders from followers.

When racing finally commenced at the Red Bull Ring in Austria, Mercedes immediately asserted its customary advantage. Lewis Hamilton and Valtteri Bottas picked up where they had left off, with the team's consistent speed advantage evident across various track layouts. Red Bull's and Ferrari's efforts to challenge the Silver Arrows fell short; Ferrari, in particular, struggled with a car that proved aerodynamically and mechanically tricky to optimize. Meanwhile, Hamilton and Bottas maintained the level of excellence expected of them, converting pole positions into victories and collecting valuable points.

The W11's reliability and performance were complemented by Hamilton's relentless focus. He approached 2020 knowing that if he clinched another Drivers' Championship, he would match Michael Schumacher's record of seven titles—an achievement once considered unattainable. With each passing Grand Prix, Hamilton steadily accumulated wins. He scored 11 victories out of 17 races in this shortened calendar, showcasing skill in all conditions: from dominant drives under sunny skies to masterful performances in wet and slippery scenarios.

Off-track, the team's culture and values came into sharper focus. Mercedes adopted a striking all-black livery for 2020, symbolizing a stand against racism and inequality. Hamilton, the first and to-date only Black driver in Formula One, vocalized his support for social justice movements, and Mercedes backed his stance, encouraging an open dialogue within the team and the sport as a whole. This synergy between sporting and social values underscored that dominance on the track did not preclude engagement with broader global issues.

The season's critical moment arrived at the Turkish Grand Prix. Held at a rain-soaked Istanbul Park in November, the race presented treacherous conditions that tested even the most skilled drivers. Hamilton, starting well behind pole position, navigated the challenge with patience and precision. By managing tires meticulously and capitalizing on opponents' errors, he charged through the field to win the race. With that victory, he secured his seventh Drivers' Championship, equaling Schumacher's long-standing record. On the team side, Mercedes had already clinched its seventh consecutive Constructors' title—another unprecedented feat in the sport's history.

Despite the pandemic's disruptions, Mercedes confirmed its status as the gold standard of the turbo-hybrid era. The team's operational excellence shone through compressed weekends, evolving safety protocols, and the absence of traditional fan support. Pit stops were executed flawlessly, strategies remained nimble, and communication channels stayed robust despite the turmoil beyond the paddock.

As the shortened season concluded in Abu Dhabi, Hamilton's record-tying seventh title and Mercedes' ongoing Constructors' success offered a moment to reflect. The team had navigated a year in which predictability vanished, and yet still emerged clearly ahead. By doing so, it illustrated that resilience, innovation, and a stable internal culture could withstand any external upheaval.

Equalling Schumacher was more than a personal milestone for Hamilton; it symbolized Mercedes' integral role in elevating its driver to the highest echelons of motorsport history. The 2020 season demonstrated that the Silver Arrows could sustain dominance under the gravest of global challenges, leaving an indelible mark on the era's narrative. It reinforced the notion that true greatness endures, adapts, and thrives—no matter how unpredictable the world outside may become.

Chapter 9: The Narrow Defeat: 2021's Title Showdown

The 2021 Formula One season began with a sense that the playing field had finally narrowed. Over the winter, minor regulation changes had shifted the aerodynamic balance of the cars, and Red Bull Racing, powered by Honda's increasingly competitive engine, emerged from testing with strong form. For Mercedes, which had reigned supreme for seven consecutive seasons, this year posed a serious challenge: the rising threat of Max Verstappen.

Lewis Hamilton entered the season as a seven-time world champion, aiming to break Michael Schumacher's record by claiming his eighth title. Early races suggested that this would be no foregone conclusion. While the Mercedes W12 remained a potent machine, it lacked a decisive advantage over Red Bull's RB16B. In fact, on many circuits, Red Bull's package appeared the more planted and flexible, especially in high-speed corners.

From the season opener in Bahrain, Hamilton and Verstappen traded blows. Verstappen took pole positions with increasing frequency, while Hamilton responded with strategic nous, tire management, and racecraft honed over a decade and a half in the sport. Races turned into tactical duels, where pit stop timing, undercuts, and safety cars could tip the balance. Neither driver had the luxury of coasting to victory; every point was hard-fought.

Crucially, the intensity extended beyond raw performance. Hamilton and Verstappen found themselves wheel-to-wheel more often than not, leading to several incidents that set the paddock and global viewership alight with debate. High-profile clashes at the British and Italian Grands Prix epitomized the ferocity of their rivalry. At Silverstone, contact between the two saw Verstappen crash heavily, intensifying tensions between Mercedes and Red Bull. At Monza, a slow pit stop and subsequent attempt to pass ended with Verstappen's Red Bull perched atop Hamilton's Mercedes in the gravel trap, the protective halo device preventing a more serious outcome. Such moments fueled storylines that transcended technicalities, painting the season as a narrative of two elite drivers pushing the boundaries.

Meanwhile, Valtteri Bottas continued to perform as Mercedes' dependable second driver, scoring vital points and even taking a well-deserved victory in Turkey. His contributions, along with the collective effort of the team's engineers, strategists, and mechanics, kept Mercedes in contention for the Constructors' Championship. Red Bull's Sergio Pérez also played a key role for his team, holding off Hamilton at crucial junctures, thus adding another layer of intrigue to the championship duel.

As the season approached its climax, Verstappen held the upper hand in the standings. Yet Hamilton, rejuvenated by late-season performances and upgrades to the W12, roared back into contention, winning races in Brazil, Qatar, and Saudi Arabia. The championship went to the wire—both drivers arrived at the final round in Abu Dhabi level on points, a scenario not seen in decades.

The Abu Dhabi Grand Prix seemed set to favor Hamilton after he took the lead early. He controlled the race pace, building what looked like a decisive advantage. However, a late crash triggered a safety car period, bunching the field. Red Bull gambled by putting Verstappen onto fresh, soft tires, while Hamilton stayed out on older rubber to maintain track position. In the closing laps, a controversial decision by race control allowed only certain lapped cars between Hamilton and Verstappen to unlap themselves, setting up a one-lap shootout. On that final lap, Verstappen, with superior grip, overtook Hamilton, clinching his first Drivers' Championship. The outcome generated immediate and intense debate, as many questioned the application of race procedures and stewarding calls. The FIA later announced an internal review of the handling of the safety car protocol.

For Mercedes, the final moments at Yas Marina were a bitter blow. Hamilton had been poised to capture that record-breaking eighth title, only to see it slip away in extraordinary and contentious circumstances. Yet, despite the disappointment on the Drivers' side, Mercedes had secured its eighth consecutive Constructors' Championship—an unprecedented achievement that underscored the team's enduring excellence. Every engineer, strategist, and mechanic had played a role in winning the season-long battle against Red Bull, ensuring that the Silver Arrows' legacy continued in the record books.

The 2021 finale highlighted the razor-thin margins that define success and failure in Formula One. While Hamilton's quest for the outright record in Drivers' titles would have to wait, Mercedes could still hold its head high. The team's performance had not diminished, even in the face of the fiercest challenge yet. And so, as the paddock dispersed into the winter, the sport found itself reflecting on a year of extraordinary tension, brilliant drives, and a conclusion that would be discussed for years to come. In that atmosphere of reflection, Mercedes knew that the fire of competition still burned brightly, and the motivation to return stronger remained undimmed.

Chapter 10: Struggling with Change: The 2022 Regulations Shift

By the time the 2022 season rolled around, Formula One had embarked on one of its most significant overhauls in decades. A radical new set of technical regulations introduced ground-effect aerodynamics, simplified front and rear wings, and larger, heavier 18-inch Pirelli tires. The aim was to promote closer racing, reduce aerodynamic turbulence, and level the competitive landscape. For a team that had reigned supreme for eight consecutive Constructors' Championships, these changes posed both a challenge and an opportunity. Mercedes, accustomed to mastering every rule set it encountered, now had to confront an aerodynamic puzzle unlike anything it had seen in the turbo-hybrid era.

The Mercedes W13 emerged from the factory bearing a slimmed-down sidepod concept and radical packaging choices. Onlookers noted its distinctive "zero sidepod" design—a bold interpretation of the new rules. Yet, what looked revolutionary on paper did not translate seamlessly to on-track performance. Almost immediately, the W13 exhibited pronounced "porpoising," a phenomenon where the car oscillated rapidly at high speeds due to the ground-effect aerodynamics. This violent bouncing not only unsettled the drivers, Lewis Hamilton and the newly signed George Russell, but also inhibited the team's ability to extract the full potential of the car's downforce. Engineers scrambled to find solutions, but the complexity of the new rules meant that each attempted fix had its own knock-on effects.

As the season began, it was apparent that Mercedes was on the back foot. Rival teams, particularly Red Bull and Ferrari, had interpreted the regulations more effectively, delivering cars that were both fast and more stable. For the first time in many years, Mercedes found itself consistently outpaced on Saturdays and forced into damage-limitation mode on Sundays. Instead of chasing victories, Hamilton and Russell were often focused on salvaging podiums or battling in the thick of the midfield at certain circuits.

The struggles were not for lack of effort. Mercedes' technical team, under the guidance of Toto Wolff, James Allison, and Mike Elliott, worked tirelessly to understand and mitigate porpoising. They revised the floor design, trialed new suspension settings, and introduced incremental updates to the aerodynamic package. Improvements came, but they arrived slowly, and always with the caveat that each gain in stability or downforce risked re-introducing unwanted bouncing on another type of circuit. Patience and methodical problem-solving became the order of the day, a stark contrast to the effortless dominance of previous seasons.

While George Russell's consistent driving and calm approach often netted solid results, the campaign tested Hamilton in a way he had not experienced at Mercedes. Since joining the team in 2013, he had always managed at least one Grand Prix victory per season—a remarkable streak that extended all the way back to his rookie year with McLaren in 2007. In 2022, however, that streak came under threat. Week after week, Hamilton confronted handling instabilities and an inherent pace deficit that left him more concerned with understanding the car's quirks than fighting at the very sharp end of the grid.

Still, there were glimpses of progress. By mid-season, Mercedes had reduced the severity of porpoising and began to inch closer to the front runners. Podiums became more frequent, and the W13's performance stabilized enough to allow for more aggressive strategies. In the latter part of the year, the team nearly tasted victory a handful of times. The breakthrough came at the São Paulo Grand Prix in Brazil, where Russell seized the moment. With strong qualifying performances and smart tire management, he claimed his maiden Formula One victory. Hamilton, finishing second, secured the team's only one-two result of the season. It was a bright spot in an otherwise challenging year, proving that Mercedes still had the fundamental capabilities to run at the front when conditions aligned.

Despite this late-season resurgence, Hamilton's record of winning at least once in every season of his career came to an end. He closed 2022 without a single Grand Prix victory, marking a clear departure from the norm. Additionally, Mercedes slipped in the Constructors' standings, finishing third behind Red Bull and Ferrari. After years of championship after championship, the team now faced the realization that absolute supremacy could never be guaranteed.

The 2022 season served as a humbling reminder that dominance in Formula One hinges on the ability to adapt to shifting ground. The ground-effect regulations had reset the field to some extent, exposing weaknesses and challenging preconceived notions. For Mercedes, it underscored the importance of open-minded engineering, long-term development plans, and resilience in the face of setbacks. The team departed 2022 not as the undisputed leader, but as a formidable contender forced back into the role of a hunter rather than the hunted.

As the paddock packed up for the winter, Mercedes carried its hard-earned lessons forward. The painstaking work done to understand porpoising and aerodynamic sensitivity would inform the next steps. Even if the W13 never achieved the lofty heights of its predecessors, it had strengthened the team's resolve. After so many years at the top, struggling with change proved that the measure of a great racing outfit is not just how they win, but how they respond when winning doesn't come so easily.

Chapter 11: Partial Recovery and Ongoing Challenges: The 2023 Season

In the wake of a difficult 2022 campaign, Mercedes approached the 2023 season determined to bridge the gap to the front of the field. The regulation reset the year before had exposed fundamental issues in the W13's design philosophy, particularly around ground-effect aerodynamics. For 2023, the team rolled out the W14—a carefully refined machine that incorporated lessons learned from its predecessor. Engineers targeted a more stable aerodynamic platform, aiming to reduce porpoising and achieve a broader performance window across the calendar's diverse circuits.

Early on, it became evident that the W14 represented a step forward. Porpoising was diminished to manageable levels, and the car's balance improved, allowing Lewis Hamilton and George Russell to push more confidently in qualifying and race conditions. The days of struggling just to keep pace with the front runners seemed a step closer to the past. Yet, Mercedes had to confront a new reality: the competition had not stood still. Red Bull, powered by its RB19, set a blistering standard at the front, while Aston Martin's surge and Ferrari's ongoing presence meant there were more contenders vying for top results than ever.

Throughout the opening rounds, Mercedes encountered a performance ceiling that limited their fight for outright victories. While the W14 was more predictable and consistent than the W13, it lacked the raw downforce and efficiency that the top teams enjoyed. Initially, the team's leadership, including Toto Wolff and technical personnel, spoke candidly about rethinking certain design choices. Recognizing that the zero-sidepod concept—once deemed revolutionary—offered diminishing returns, they introduced an extensive upgrade package mid-season to realign the car's concept with a more conventional approach. Small aerodynamic refinements, new sidepod profiles, and revised floor geometry all aimed to unlock latent potential.

The improvements, although incremental, yielded tangible progress. Hamilton and Russell became regular podium threats, consistently finishing in the top three or four positions. Tracks that emphasized tire management and efficient downforce distribution were where the W14 shone brightest. For Hamilton, these steady podiums underscored his enduring ability to extract the maximum from any car, while Russell's calm methodology and adaptability ensured that the team maintained a strong presence in the Constructors' Championship standings.

Yet, despite these encouraging signs, a race victory remained elusive. Mercedes found themselves regularly looking at the back of Red Bull's dominant RB19 and occasionally grappling with Aston Martin's and McLaren's improved machines. Grand Prix weekends turned into exercises in precision: qualifying well, managing strategy to protect against undercuts and overcuts, and capitalizing on any rival misfortune. While these efforts often translated into solid points, they never quite delivered the top-step celebrations that had once been so familiar.

By the mid-to-late stages of the season, Mercedes had established itself as a firm podium contender and, in some races, the second-fastest team on the grid. This stability enabled them to solidify second place in the Constructors' Championship by the year's end. Given the difficulties of 2022, claiming the runner-up spot was not a trivial achievement. It represented a partial recovery, a sign that the team's development path was headed in the right direction—even if it had not fully closed the performance deficit to Red Bull.

For Hamilton, the season marked another year without a victory—something unimaginable during the team's dominant streak just a few years earlier. Yet, his string of podium finishes, including multiple second-place results (notably in Australia, Spain, Canada, Great Britain, and the Netherlands), showed that he remained a formidable force. Russell, too, continued to mature, bringing consistency and insightful feedback that guided ongoing development. Both drivers embraced the challenge of rebuilding Mercedes' competitive edge from a position that demanded persistence rather than comfort.

As the 2023 season drew to a close, the team could reflect on what it had accomplished and where it still fell short. Mercedes had transformed a troubled concept into a more competitive package, learned from failures, and regained some lost ground. Although the top step of the podium remained out of reach, the year's work laid a stronger foundation for future campaigns. Mercedes now knew the direction to pursue and the pitfalls to avoid. Acknowledging that Formula One's landscape had changed and that dominance could not be assumed, the team left 2023 committed to turning incremental gains into decisive strengths for the seasons ahead.

Ultimately, 2023 was about rebuilding credibility, refining concepts, and reconfirming Mercedes' status as a top-tier contender even when not at its absolute peak. Armed with hard-earned insights, the team prepared to face whatever challenges the next chapter of the sport would bring.

Chapter 12: Last Season Together: The 2024 Season

In 2024, Lewis Hamilton embarked on his final season with Mercedes, a partnership that had yielded six World Championships and countless records. The year was marked by both challenges on the track and the emotional weight of an impending farewell.

Hamilton's 2024 Season

The season began under the shadow of Hamilton's announcement that he would join Ferrari in 2025, a decision that surprised many and added a layer of complexity to his final year with Mercedes. Hamilton later admitted he had "massively underestimated" the difficulty of navigating his departure, struggling to inform the team and manage his emotions throughout the season.

On the track, Hamilton demonstrated his enduring brilliance with victories at the British Grand Prix, ending a 945-day winless streak, and at the Belgian Grand Prix. These moments served as reminders of his remarkable skill and determination. However, the season also presented struggles, such as a disappointing qualifying session in Abu Dhabi, where a dislodged bollard compromised his final race weekend with Mercedes.

Ultimately, Hamilton finished the season seventh in the Driver Standings, marking his lowest career finish and the first time he ended a season outside the top six. Despite these challenges, Hamilton expressed profound gratitude towards Mercedes, reflecting on their shared triumphs and the deep bonds formed over 12 years.

The Heartfelt Separation and Move to Ferrari for the 2025 Season

The decision to leave Mercedes for Ferrari was not made lightly. Hamilton's move was driven by the allure of Ferrari's unparalleled prestige and a desire for a new challenge in the twilight of his illustrious career. The transition was emotionally taxing, as he grappled with the reality of departing a team that had become like family. Informing team principal Toto Wolff of his decision was particularly difficult, given their close and enduring relationship.

The announcement had significant implications for both teams. Mercedes had to pivot their future plans, promoting young talent Andrea Kimi Antonelli to fill Hamilton's seat. Wolff acknowledged the challenge of replacing a driver of Hamilton's caliber but remained focused on the team's future prospects.

As Hamilton prepared to don the iconic red of Ferrari, the Scuderia extended a symbolic gesture of integration and respect by granting him residence in Enzo Ferrari's former home—an honor previously bestowed only upon Michael Schumacher. This move underscored the significance of his arrival and Ferrari's commitment to their newest driver.

Hamilton's final season with Mercedes was a poignant conclusion to one of the most successful partnerships in Formula 1 history. As he transitions to Ferrari, he carries with him the legacy of a remarkable journey and the anticipation of new challenges with the Scuderia.

Chapter 13: Technical Innovations and Car Development

Throughout Lewis Hamilton's tenure at Mercedes, the team's sustained success owed much to a relentless drive for technical innovation. From the moment Hamilton joined in 2013, Mercedes approached car development with a clear philosophy: pair intelligent engineering solutions with meticulous refinement to stay ahead of evolving regulations and a fiercely competitive field. This forward-thinking attitude shaped both their aerodynamic concepts and the power units that powered them to unprecedented heights.

Engineering Milestones in the Hybrid Era
When Formula One introduced the turbo-hybrid power units in 2014, Mercedes was ready. Their work began years earlier at Brixworth, where Mercedes High Performance Powertrains (HPP) laid the groundwork for a class-leading engine. At the heart of their approach was the decision to separate the turbocharger's turbine and compressor units, placing them at opposite ends of the internal combustion engine's V6 block. This ingenious layout improved packaging, reduced turbo lag, and contributed to better overall efficiency. The result: from the start of the hybrid era, the Mercedes

power unit delivered a potent blend of horsepower and drivability, providing a decisive advantage on track.

Year by year, Mercedes refined these hybrid systems. Their engineers optimized combustion efficiency, integrated advanced materials to reduce friction and weight, and worked with partner Petronas to develop fuels and lubricants that extracted maximum performance and reliability. Continuous improvements to the MGU-K and MGU-H—components capturing and deploying electrical energy—enhanced energy recovery and deployment, enabling drivers to rely on consistent power boosts throughout a race lap. Over time, incremental gains added up. While rivals often struggled with reliability or thermal management, Mercedes maintained a reputation for bulletproof engines that allowed full-throttle racing without compromising endurance.

Aerodynamic Prowess and Adaptation
On the aerodynamic front, Mercedes consistently responded to changing regulations with assured technical leadership. When the team first came to prominence in 2013, they built upon a solid foundation and used the following years to master new aerodynamic philosophies. The transition to the hybrid era coincided with rules limiting fuel flow and engine power, making efficient aerodynamics even more

critical. The championship-winning W05 of 2014 exemplified a well-balanced approach: a car that complemented the power unit's superiority with low drag and stable downforce.

In subsequent seasons, Mercedes overcame a series of aerodynamic challenges. The 2017 regulations introduced wider cars, larger wings, and a significant leap in downforce. While this shift allowed rivals to gain ground, Mercedes adapted quickly, ensuring their W08, W09, and W10 retained strong baseline performance. They fine-tuned suspension geometry, honed underbody aero, and incorporated complex turning vanes, bargeboards, and brake duct winglets—all thoroughly tested in the wind tunnel and verified through track data. Such refinements kept Mercedes at or near the front even as Ferrari and Red Bull intensified their efforts.

Not all innovations were visible at a glance. Intensive work on mechanical setup and tire management ensured that aerodynamic balance did not come at the cost of excessive tire wear. This synergy between aero and mechanical grip allowed Hamilton and his teammates to maintain competitive lap times deep into a race stint, strategically outfoxing competitors who struggled to preserve their tires.

Groundbreaking Concepts and Systems
Some Mercedes innovations became talking points in the paddock. In 2020, the team introduced the Dual Axis Steering (DAS) system on the W11, enabling the drivers to adjust the front wheel toe angle by pushing or pulling the steering wheel along its axis. This clever mechanism helped optimize tire temperatures on straights and improved turn-in response, providing a subtle but tangible advantage. Although DAS was short-lived—outlawed by new regulations the following year—it exemplified the team's willingness to explore creative solutions within the rulebook.

Over time, Mercedes also experimented with novel aerodynamic architectures. The W12 and W13 showcased a "zero-sidepod" concept, a visually striking approach that aimed to reduce drag and open up new airflow pathways. While the W13 struggled in 2022 due to porpoising issues and the inherent complexities of the ground-effect regulations, these attempts represented the team's constant search for fresh ideas. Even when concepts did not yield immediate success, the knowledge gained informed future developments, ensuring that no experiment was truly wasted.

Sustained Advancement and Continual Learning
At the core of Mercedes' technical philosophy was a

culture of continuous improvement. The flow of data between track and factory was seamless and swift. Engineers pored over telemetry, wind tunnel results, and computational fluid dynamics simulations, seeking areas to trim hundredths of a second. The close integration between the engine department at Brixworth and the chassis team at Brackley ensured that power unit and aerodynamics evolved in tandem, each development feeding into the other.

This ecosystem thrived on collaboration. Lessons learned from successes—such as the highly efficient 2014 power unit or the aerodynamically near-flawless W11 of 2020—fed directly into tackling new challenges. When the 2022 regulations triggered porpoising and unsettled handling, Mercedes' technical staff regrouped, re-analyzed their data, and steadily introduced refinements. Although this process took time, by 2023 they had built a more stable baseline, ready to evolve further.

A Legacy of Innovation
Mercedes' technical journey during Hamilton's stay at the team stands as a masterclass in adaptability and innovation. Across shifting rules, aerodynamic philosophies, and engineering paradigms, the team's emphasis on robust design principles and meticulous refinement proved decisive. The hybrid power units

they created set the standard for performance and reliability, and their aerodynamic concepts—whether traditional or radical—consistently pushed the envelope.

This legacy, forged through championship glory and technical risk-taking, contributed to Hamilton's success on the track. In the end, it was the sum of countless details: split-turbine layouts, precision-molded carbon fiber components, minutely tuned suspension elements, and experimental concepts that sometimes bent, but never broke, the rules. Taken together, these engineering milestones tell the story of how a technical powerhouse at Mercedes drove Hamilton forward, season by season, into the annals of Formula One history.

Chapter 14: Team Management and Strategy

At the heart of Mercedes' dominance during Lewis Hamilton's tenure was a management structure that balanced strong leadership, open communication, and a commitment to continuous improvement. Under the stewardship of Toto Wolff, who joined Mercedes in an executive role in 2013, the team cultivated a collaborative atmosphere that emphasized accountability without blame, data-driven strategy calls, and long-term thinking over short-term fixes. This approach allowed Mercedes to maintain a high standard even as rivals closed in, regulations shifted, and internal challenges arose.

Leadership Under Toto Wolff

A former racing driver turned investor and team principal, Toto Wolff quickly became the public face and strategic mastermind of Mercedes-AMG Petronas Formula One Team. Taking on increasing responsibilities after the departure of Ross Brawn at the end of 2013, Wolff worked closely with Niki Lauda (as Non-Executive Chairman until his passing in 2019) to shape a stable, forward-looking organization. Wolff's leadership style blended clear decision-making with a willingness to listen—he was known to encourage input from all levels of the team, from aerodynamicists and engine technicians to trackside engineers and strategists.

Central to Wolff's philosophy was the concept of a no-blame culture. Mistakes were not punished; instead, they were dissected methodically to ensure the team learned from them and came back stronger. If a pit stop went awry, or a strategic call backfired, the post-race debriefs would not turn into finger-pointing sessions. Instead, key data would be reviewed, simulations re-run, and alternative scenarios explored. This approach fostered trust and loyalty, ensuring that everyone involved felt safe to propose ideas or highlight problems without fear of reprisal.

Another hallmark of Wolff's management was talent retention and development. Key technical figures—engineers, strategists, and designers—were given long-term support and the resources needed to pursue ambitious solutions. This long-view strategy stabilized the organization and cultivated in-house expertise. As rival teams frequently shuffled their personnel, Mercedes maintained a core group of individuals who understood the team's ethos, processes, and objectives. Such continuity was a cornerstone of Mercedes' success, helping to integrate new technologies and respond quickly to regulatory changes.

Documented Strategic Calls and Management Philosophies

Mercedes' strategic decisions on race weekends were shaped by data-driven analysis, sophisticated simulation tools, and a disciplined command structure on the pit wall. The team's chief strategist, initially James Vowles during Hamilton's prime years, played a pivotal role in making split-second calls on tire changes, fueling strategies, and responding to safety cars. Communications over team radio, which became public after races, revealed a calm, calculated tone—evidence of a well-prepared strategy unit working through possible outcomes before finalizing instructions.

One of the most documented aspects of Mercedes' strategy was their emphasis on minimizing risk while maintaining track position. With a generally faster car at their disposal during the height of their dominance, they often prioritized reliability and consistent scoring over gambler's strategies. This conservative yet calculated approach won them multiple titles. Still, when circumstances demanded, Mercedes showed the flexibility to adapt. For example, mid-race reactions to sudden weather changes, safety cars, or red flag interruptions often saw them anticipate rivals' moves, preserving advantages or mitigating potential losses.

The team's internal rivalry management also showcased their strategic philosophy. During the Hamilton-Rosberg years (2013–2016), tensions flared as both drivers fought for championships. Rather than impose rigid team orders at the outset, Mercedes set clear expectations: both drivers were free to race but must avoid compromising the team's results. After incidents such as the collision in Spain 2016, Wolff and Lauda convened with the drivers, emphasizing accountability and mutual respect. This measured approach allowed intense competition without fracturing the team spirit.

In subsequent years, as Hamilton faced growing competition from other teams—Ferrari's Sebastian Vettel and later Red Bull's Max Verstappen—Mercedes' strategy focused on proactive adaptation. Documented examples included carefully timed pit stops under virtual safety cars, long-run stints to force rivals into unfavorable strategies, and agreeing to strategic compromises between drivers to secure the best possible combined result. The overarching logic always involved using their deep knowledge base, simulation capabilities, and established protocols to chart the least risky path to victory.

Moreover, after the controversial finale of the 2021 season, Mercedes' response mirrored their core management principles. Though deeply disappointed by the outcome—which hinged on a late-race interpretation of safety car procedures—Wolff publicly balanced frustration with calm resolve. Internally, the team conducted thorough reviews, accepting the situation and focusing on constructive steps for the future rather than dwelling on grievances. This ability to rebound from setbacks, supported by well-established debrief processes and a no-blame culture, enabled Mercedes to remain competitive even when not in outright control.

Building a Winning Culture
Ultimately, Mercedes' management and strategic philosophy revolved around clarity, mutual respect, and a willingness to evolve. While Wolff's presence at the helm provided strong and consistent leadership, the collective decision-making structure ensured that no single voice dominated. It was the combination of top-level vision and grassroots contribution that produced an environment where every team member understood their role, felt valued, and strove to exceed expectations.

This approach allowed Mercedes to set and maintain high standards, not just in terms of raw speed or engineering marvels, but also in the subtleties of on-track decision-making and off-track human resource management. The team navigated rule changes, intense rivalry, and external controversies without losing its identity or mission. In doing so, it became not only a championship-winning machine but also a model of how modern motorsport teams could function at their best—organized, analytical, humane, and resilient.

Chapter 15: Key Rivalries and Partnerships

From the moment Lewis Hamilton joined Mercedes in 2013, the dynamics of his working relationships—both inside and outside the team—shaped his journey. While Mercedes' engineering prowess and strategic acumen provided the foundation, the tensions, alliances, and respect forged on track and in the garage were equally integral to his narrative. Over the years, Hamilton's interactions with teammates Nico Rosberg, Valtteri Bottas, and George Russell, as well as his intense battles against top rivals Sebastian Vettel and Max Verstappen, defined an era of high-stakes competition at the pinnacle of Formula One.

Teammate Dynamics: Rosberg, Bottas, and Russell
When Hamilton arrived at Mercedes, he stepped into a garage occupied by Nico Rosberg, who had driven for the team since its reformation in 2010. Initially, the atmosphere was cordial: two drivers of similar age, with past ties in junior formulas, united under a team striving to reach the front. However, once the turbo-hybrid era began in 2014 and Mercedes delivered a car capable of dominating the field, the stakes soared. Two highly competitive drivers were now contending for race wins and world championships with the same machinery. In

2014 and 2015, Hamilton clinched the title, but Rosberg never relented. By 2016, the tension had intensified. High-profile incidents—like their collision in Spain—demonstrated how fine the line was between pushing each other and risking the team's results. That year, Rosberg outlasted Hamilton to claim the Drivers' Championship by a narrow margin, retiring shortly afterward and leaving behind a legacy of fierce but balanced rivalry. The Hamilton-Rosberg dynamic became a case study in managing intra-team competition at the highest level.

Valtteri Bottas joined Mercedes in 2017, stepping into the seat vacated by Rosberg. Compared to the combative earlier years, the Hamilton-Bottas partnership was calmer and more cooperative. Bottas earned multiple victories and often supported Hamilton's title bids by taking valuable points off rival teams. There were no major flashpoints comparable to the Rosberg era. Instead, Bottas' consistent professionalism, adaptability, and willingness to play the team game allowed the partnership to flourish. While Bottas aimed to challenge Hamilton, the balance of power typically remained in Hamilton's favor. Still, Bottas' contributions—including key poles, race wins, and strategic defense—were instrumental in maintaining Mercedes' run of Constructors' Championships.

In 2022, the team brought in another new teammate: George Russell. Younger and eager to prove himself, Russell arrived with experience at Williams and strong junior credentials. He quickly adapted, consistently qualifying and finishing near the front despite a challenging car. Russell's calm feedback and analytical approach mirrored some of Bottas' traits, but he brought a fresh perspective and posed a more direct on-track challenge to Hamilton's supremacy. While the W13's and W14's limitations prevented a full-blown title fight, Hamilton and Russell collaborated to develop the car and push the team forward. The first full seasons together saw a respectful rapport, with Russell's maiden win in 2022 and strong results in 2023 indicating that he could be a genuine long-term rival within the same garage.

On-Track Battles with Vettel, Verstappen, and Others

Beyond the Mercedes garage, Hamilton's primary adversaries shifted with the competitive tides. In the earlier years of the hybrid era, Sebastian Vettel and Ferrari emerged as the main threats. From 2017 to 2019, Hamilton and Vettel frequently went wheel-to-wheel. Vettel's Ferrari, especially in 2017 and the first half of 2018, occasionally held the upper hand. These seasons featured close title fights, with both Hamilton and Vettel trading victories and psychological blows. Highlights included Hamilton's determined driving at circuits like Spa and Monza, where the tension between the two champions created compelling narratives. Ultimately, Hamilton and Mercedes proved more consistent over the full campaign, fending off Vettel's challenges and demonstrating superior strategic adaptability. While their rivalry was intense, it was generally characterized by mutual respect, informed by their status as multiple world champions.

In 2021, a new contender rose to the forefront: Max Verstappen. Younger, aggressive, and driving for a rapidly improving Red Bull, Verstappen challenged Hamilton in a season-long duel that proved one of the most dramatic in modern F1 history. The pair's battles were often decided by the smallest margins. Their encounters at Silverstone, Monza, and the controversial finale in Abu Dhabi exemplified just how evenly matched and fierce their rivalry had become. Verstappen's unapologetic racing style pushed Hamilton out of his comfort zone, forcing him and the team to make delicate strategic decisions and respond to pressure unfamiliar in the hybrid era. Their rivalry was more than a clash of talent; it was a generational battle, pitting a seasoned champion against an emerging star. Although Hamilton narrowly missed the Drivers' title in 2021, the season cemented both drivers' reputations as outstanding competitors.

Hamilton also faced secondary challenges from other talented drivers over the years. Daniel Ricciardo, Charles Leclerc, and Fernando Alonso (in his later return to competitive form) occasionally served as key rivals in individual races. While these engagements were less sustained than his battles with Vettel or Verstappen, they demonstrated Hamilton's ability to adapt his racecraft to any challenger, whether fighting an experienced world champion or a rising newcomer.

A Legacy of Competition and Respect
Over the span of a decade at Mercedes, Hamilton's relationships with teammates and key rivals traced an arc of conflict, camaraderie, and evolution. Inside the garage, the team navigated delicate balances between equality and competitiveness. Outside it, Hamilton faced off against some of the finest drivers of his era, forging rivalries that tested his skill, resilience, and mindset.

The intensity of the Hamilton-Rosberg fights, the solidarity and cooperation with Bottas, the synergy and rebuilding phase alongside Russell, the pitched battles against Vettel's Ferrari, and the explosive showdown with Verstappen's Red Bull—together, these interactions shaped an epoch in Formula One. They underscored that greatness in racing comes not just from outright speed, but from how a driver engages with those around him: respecting teammates while challenging them, pushing rivals without disregarding safety and fairness, and ultimately leaving a legacy defined by honor, determination, and relentless pursuit of excellence.

Chapter 16: Records and Milestones

Over the span of his Mercedes career, Lewis Hamilton built a statistical resume that redefined the meaning of success in Formula One. Arriving at the team in 2013 as a single-time world champion, Hamilton departed the end of the 2023 season with records and milestones that had previously seemed out of reach. Through official race results, FIA statistics, and documented facts, Hamilton's numbers tell a story of sustained dominance, adaptability, and relentless pursuit of excellence.

Wins, Poles, and Podiums: Shattering the Record Books

One of the clearest measures of Hamilton's impact during his time at Mercedes is his tally of race victories. Although he entered the team with 21 wins (all secured at McLaren), the era from 2013 to 2023 saw him accumulate a total that not only surpassed the longstanding record held by Michael Schumacher but also pushed beyond the century mark—an unprecedented feat. By the close of the 2023 season, Hamilton stood at 103 Grand Prix victories. The overwhelming majority of these came with Mercedes power, reflecting the unparalleled synergy between driver and team during the turbo-hybrid era.

Qualifying prowess was another arena in which Hamilton excelled. Securing pole positions at a wide array of circuits—from power-dependent tracks like Monza to twisty layouts like Hungary—he consistently converted one-lap pace into an art form. His pole position record, like his win total, soared past Schumacher's previous benchmark. By late 2021, Hamilton had already crossed the 100-pole threshold, demonstrating consistent qualifying excellence year after year. As of the end of 2023, he held 103 pole positions, a testament to his enduring speed over a single lap.

Podiums serve as a broader indicator of sustained performance, and here too, Hamilton's numbers were formidable. With more than 190 podium finishes in total, he secured not just race victories and front-row starts but also a remarkable level of consistency. Week in and week out, through regulation changes and shifting competitive orders, Hamilton found ways to reach the top three, often salvaging results from difficult weekends and capitalizing on the W05, W06, W07, and so forth, right up to the W14, to achieve a steady stream of trophies.

Championships and Other Landmark Achievements
Central to Hamilton's legacy at Mercedes are the six

Drivers' Championships he secured with the team, added to his initial title from McLaren. Claiming titles in 2014, 2015, 2017, 2018, 2019, and 2020, he elevated his total to seven—matching Schumacher's record. These championships were earned under varying conditions: from the dominance of early hybrid-era campaigns to the intense, season-long battles against Sebastian Vettel, and the pandemic-shortened 2020 year where Hamilton showcased resilience and adaptability.

Over his Mercedes tenure, Hamilton also became the first driver in Formula One history to reach 100 Grand Prix victories and 100 pole positions, shattering a psychological barrier that once seemed insurmountable. His 100th win came at the 2021 Russian Grand Prix, while his 100th pole had been secured earlier that year in Spain. Both moments were greeted as milestones not just for the driver, but for the sport itself, signaling a new standard of achievement.

Statistical Versatility and Longevity
Beyond the headline records, Hamilton's statistics reveal a remarkable longevity. He consistently scored championship points in the overwhelming majority of races, a result of both his driving skill and Mercedes' reliable engineering. His streak of winning at least one race in every season of his career until 2022 highlighted an unparalleled ability to adapt to evolving cars, tires,

regulations, and competition. Though that streak ended in 2022, the broader pattern of frequent podiums and top-five finishes underlined his sustained competitiveness over a full decade.

Hamilton's success was not confined to specific track types or conditions. His wins spanned continents, climates, and configurations—he collected victories on street circuits like Monaco and Singapore, high-speed temples of speed like Monza and Silverstone, and technical layouts like Suzuka and Interlagos. Rain or shine, Hamilton's statistical record shows an ability to thrive anywhere.

A High-Water Mark in Motorsport History
All these numbers combined set a benchmark that future generations of drivers will chase. While records in sport are inevitably broken over time, the scale and breadth of Hamilton's achievements with Mercedes have established a new reference point. His totals of wins, poles, and podiums stand as evidence of what is possible when a driver's talent, a team's technical ingenuity, and a shared hunger for success align perfectly.

Additionally, the documented data serves an archival purpose. Each victory, pole, podium, and fastest lap is recorded by the FIA and disseminated through official statistics and reputable motorsport databases. This ensures that Hamilton's accomplishments are preserved and accessible, allowing fans, historians, and competitors to understand the magnitude of his performance within its proper factual context.

Defining an Era Through Numbers

Ultimately, the statistical portrait of Hamilton's Mercedes era is one of extraordinary accomplishment. He rose from a single world title holder in 2013 to a driver who rewrote the record books, normalized the extraordinary, and set a standard for consistency and excellence in the most competitive era of Formula One. While the sport marches on and future talents will strive to emulate or surpass these feats, Hamilton's records remain a defining feature of his legacy, one etched into the annals of motorsport history with stark and undeniable clarity.

Chapter 17: Cultural Impact and Media Response

Over the course of Lewis Hamilton's career at Mercedes, the cultural footprint of both driver and team expanded well beyond the traditional Formula One fanbase. Documented shifts in global viewership, social media engagement, and media commentary revealed that Hamilton's success, combined with Mercedes' dominance, influenced how the sport was perceived and followed worldwide. Drawing on official viewership data, media reports, and recognized awards, this chapter chronicles the factual trends that emerged from Hamilton's tenure.

Global Fan Engagement and Viewership Trends

Formula One's global television audience has long been substantial, consistently reaching hundreds of millions of viewers annually. According to official Formula One Management (FOM) reports, the hybrid era (beginning in 2014) saw stable and in some years increased television and digital viewership, due in part to the intense battles and evolving narratives at the front of the grid. While several factors contributed—such as improved broadcasting deals, increased digital presence, and racing unpredictability—Hamilton's prominent role as a multiple-time champion and a recognizable figure across continents played an indirect role in keeping the sport in the public eye.

During key title-deciding seasons, especially high-profile duels in 2016, 2018, 2019, and particularly the 2021 season against Max Verstappen, Formula One reported rising interest. The final race of 2021, in which Hamilton fought for an eighth championship, drew an official global viewership of over 108 million viewers—one of the largest live Formula One audiences ever recorded. While that season's outcome was contested on multiple levels, the numbers highlighted that storylines involving Hamilton's quest for records and Mercedes' dominance were central to engaging audiences.

In addition to traditional broadcast coverage, the sport's social media platforms—Twitter, Instagram, Facebook, and later TikTok—recorded notable growth during Hamilton's peak years. Formula One's digital engagement data, regularly published by the organization, showed surges in follower counts and video views throughout the late 2010s and early 2020s. While this increase cannot be attributed to one individual alone, prominent figures like Hamilton, featured regularly in official F1 content and behind-the-scenes media, contributed to heightened interest and fan conversations online.

In North America, historically a more limited market for Formula One, various factors—including broader marketing strategies, expanded coverage through ESPN in the United States, and the popularity of documentary series such as Netflix's "Drive to Survive"—coincided with Hamilton's record-breaking years. Although official sources do not isolate viewership spikes by driver, commentators and media analysts frequently cited Hamilton's global profile as one of the drivers of F1's increased resonance with younger and more diverse audiences, especially in markets where the sport had previously struggled to gain traction.

Documented Media Commentary and Public Perception Shifts

Media outlets across the world, from Europe's motorsport-focused press to general-interest publications in Asia, Africa, and the Americas, covered Hamilton's achievements extensively. He appeared frequently on mainstream platforms, including the BBC, CNN, and major European newspapers, gaining recognition not only for his on-track success but also for his off-track persona. Official acknowledgments of his influence included Time magazine naming him one of the world's 100 most influential people in 2020.

Commentary in respected motorsport publications such as Autosport and Motorsport.com often highlighted Hamilton's unique place in the sport's history, particularly after surpassing Michael Schumacher's long-standing records for wins and pole positions. Journalists documented how Hamilton's achievements at Mercedes prompted comparisons that once seemed inconceivable, and his pursuit of statistical landmarks became a recurring narrative that broadened public interest.

Additionally, Hamilton's engagement with social issues—well-documented through official press conferences, social media posts, and verified interviews—shaped his public perception. Media coverage of Hamilton's advocacy for diversity, inclusion, and social justice underscored how a top-tier driver at a leading team could transcend the sport's traditional boundaries. This shift in discourse did not detract from coverage of the racing itself; rather, it enriched public understanding of the champion as a multifaceted individual. As a result, journalists frequently credited Hamilton with helping Formula One appear more accessible and relevant to a global, 21st-century audience.

Reflected Legacy in Popular Culture

By the late 2010s and early 2020s, Hamilton's presence, along with Mercedes' sustained success, was woven into global sporting culture. His recurring victories at iconic circuits, appearances at award ceremonies—such as receiving the BBC Sports Personality of the Year multiple times—and endorsement by reputable outlets consolidated a narrative of ongoing excellence. Media coverage documented not only his race-by-race performance but also the long-term impact of his record-breaking journey, ensuring that his accomplishments with Mercedes were recognized beyond the traditional motorsport community.

While precise audience metrics and public opinion data are complex and influenced by many variables, the official facts are clear: Formula One's reach remained robust, social platforms saw growth, and authoritative media consistently spotlighted Hamilton's role in elevating storylines that captivated fans worldwide. Over the course of a decade with Mercedes, these measurable trends confirmed that Hamilton's and the team's cultural resonance was no accident. It reflected a mutually reinforcing cycle where sporting achievement fueled interest, and that interest, in turn, broadened the sport's cultural footprint.

Chapter 18: Safety and Sporting Regulation Changes

During Lewis Hamilton's tenure at Mercedes (2013–2023), Formula One underwent a series of substantive safety and sporting regulation changes orchestrated by the FIA. These modifications affected everything from car construction and driver protection to race weekend procedures and officiating standards. With the FIA's role centered on improving fairness, safety, and the spectacle of racing, the period saw numerous regulations introduced or revised, many of which had direct, documented impacts on how Hamilton and all drivers approached the sport.

FIA Rule Evolutions and Their Motivations
The shift to the turbo-hybrid era in 2014 was a foundational change, primarily technical in nature, but it also prompted the FIA to refine sporting regulations to accommodate more complex power units. Over time, the regulatory body introduced cost-saving measures, limitations on engine components, and constraints on testing. While these changes aimed to ensure competitive parity and long-term financial stability, they also required teams like Mercedes to adapt their engineering and operational strategies continuously.

Beyond the technical realm, the FIA responded to high-profile incidents and evolving safety data by updating rules. This included stricter limits on track re-entry, more explicit guidelines on defending and overtaking, and adjustments to how track limits were enforced. Penalties for unsafe driving, whether involving collisions or forcing another driver off track, became more standardized. The introduction of a penalty points system on super licenses allowed the FIA to track repeated infractions. Hamilton and his peers navigated these clearer but sometimes more stringent disciplinary frameworks, knowing that accumulating penalty points could lead to race bans.

Safety Car and Virtual Safety Car Protocols
One significant sporting regulation innovation was the establishment of the Virtual Safety Car (VSC) system in 2015. Introduced after Jules Bianchi's tragic accident in 2014, the VSC allowed race control to neutralize the field without deploying the physical safety car. Under VSC conditions, drivers must adhere to a prescribed delta time, slowing the pack uniformly for short periods while marshals clear incidents safely. This measure aimed at increasing safety while minimizing race interruptions. Hamilton, like all drivers, adapted to these protocols, balancing opportunities to gain strategic advantages through quick pit stops with the strict requirement to maintain the VSC-limited pace.

The usage and interpretation of the physical safety car periods also evolved. The FIA periodically adjusted rules governing how lapped cars could unlap themselves and how restarts were conducted. By the end of Hamilton's tenure, these regulations had been revised multiple times. The controversial final-lap restart in the 2021 Abu Dhabi Grand Prix, where only some lapped cars were allowed to unlap themselves, led to scrutiny and internal reviews. Subsequent clarifications and procedural changes were made in 2022 to ensure more transparent and consistent safety car restart procedures.

The Introduction and Impact of the Halo
Perhaps the most visually and symbolically significant safety innovation during Hamilton's Mercedes era was the introduction of the halo cockpit protection device in 2018. Initially met with skepticism by some fans and drivers due to its aesthetics, the halo became a mandatory feature on all F1 cars following extensive FIA research and testing. The titanium structure was designed to protect drivers' heads from flying debris, detached wheels, and other large objects.

Documented incidents proved the halo's effectiveness. Multiple accidents during Hamilton's later seasons—such as those involving other drivers at Spa-

Francorchamps, Monza, and Silverstone—demonstrated how the halo likely prevented serious injury. The FIA publicly acknowledged cases where the halo's presence was a significant factor in safeguarding the driver. Although Hamilton never experienced a life-threatening halo-related incident himself, he and his fellow competitors universally came to accept and praise the device's protective value.

Penalties and Stewarding

The FIA continued to refine the way penalties were assessed and enforced. Dedicated steward panels, including driver stewards with racing experience, aimed to improve consistency in decision-making. Incidents were judged against increasingly detailed guidelines. Over time, the FIA published explanations for certain high-profile decisions, improving transparency. Although subjective calls were still debated publicly, the documentation of these processes allowed teams and drivers—Hamilton included—to understand the rationale behind penalties more clearly.

Data-driven stewarding also meant that radio communications, telemetry, and video replays were utilized extensively to verify what occurred during on-track clashes. This ensured that outcomes weren't based on partial evidence, and drivers adjusted their racing styles accordingly. Hamilton, known for his aggressive yet generally fair racecraft, adapted to evolving standards, often praising or critiquing the system through official media interviews. By the end of his 2023 season, the principle of consistent, data-backed stewarding was widely recognized, even if occasional controversies persisted.

Shifts in the Sporting Landscape

Other noteworthy regulatory evolutions included limitations on engine component usage per season, standardized testing reductions, tighter fuel flow rules, and periodic aerodynamic restrictions aimed at improving on-track overtaking. The FIA introduced these changes intending to prevent runaway technical advantages and to keep racing entertaining and competitive. While Mercedes thrived for years under these conditions, Hamilton repeatedly acknowledged in official press conferences that adapting to new rules kept teams and drivers alert, ensuring a dynamic sporting environment.

Over the decade Hamilton spent at Mercedes, these safety and sporting regulations shaped the conditions under which he and the entire field competed. The halo protected them from grievous harm, the VSC and safety car protocols sought to maintain fairness and order, and the evolving penalty system promoted cleaner racing. Each change, documented in official FIA briefings and resulting from ongoing consultation with teams and drivers, contributed to making the sport safer, more transparent, and ultimately fairer—principles that guided Formula One's evolution throughout Hamilton's era.

Chapter 19: Hamilton's Driving Evolution and Technical Feedback

Throughout his tenure at Mercedes, Lewis Hamilton's driving style did not remain static. Instead, it evolved in tandem with the team's machines and the ever-changing Formula One regulations. From the turbo-hybrid revolution of 2014 to the aerodynamic and tire overhauls introduced in subsequent years, Hamilton adapted his approach behind the wheel and played a key role in shaping the technical direction of the cars he drove. Team principals, technical directors, and engineers publicly acknowledged his meticulous feedback, pinpointing how his input guided setup adjustments and influenced design philosophies season after season.

Adapting to the Turbo-Hybrid Era (2014–2016)
When the V6 turbo-hybrid engines arrived in 2014, drivers faced entirely new handling characteristics—greater torque, increased complexity in energy recovery, and altered braking behavior due to the brake-by-wire systems. Hamilton embraced the challenge. Paddy Lowe, then Mercedes Executive Director (Technical), and Toto Wolff, Team Principal, noted in interviews that Hamilton's clear descriptions of how the power unit delivered torque at different throttle

positions helped the team refine engine maps and drivability settings. His feedback on pedal feel and braking stability informed adjustments to the brake-by-wire calibration, making the car more predictable under heavy deceleration.

By 2015 and 2016, Hamilton and Nico Rosberg were engaged in a fierce intra-team rivalry, and marginal gains in setup mattered more than ever. Hamilton's penchant for a strong, responsive front end to suit his aggressive corner entry style was well-documented. Engineers, speaking in race debriefs and media sessions, highlighted that Hamilton consistently requested changes to front wing angles, differential settings, and suspension tweaks that gave him the initial front-end bite he craved. This guided the team's aerodynamic refinements and mechanical setups, ensuring that the W06 and W07 cars could extract maximum performance from the Pirelli tires, even as their compounds evolved.

Responding to Aerodynamic Shifts and Tire Evolutions (2017–2019)

The major aerodynamic rule changes in 2017 produced faster, more downforce-heavy cars with wider tires. Hamilton's ability to quickly understand and exploit these new regulations stood out in official Mercedes technical briefings. James Allison, who joined as

Technical Director in 2017, publicly commended Hamilton's skill in communicating how the stiffer construction and broader footprint of the tires affected balance mid-corner. His input was critical for refining front suspension geometries and floor profiles, ensuring that the W08 and its successors maintained consistent grip over long stints.

In 2018 and 2019, Pirelli introduced thinner-tread tires to combat blistering issues. Many teams struggled initially, but Mercedes adapted more swiftly than some rivals. Post-race interviews and press conference statements from Wolff and Chief Race Engineer Andrew Shovlin frequently singled out Hamilton's nuanced feedback. He would describe precisely where the tire's temperature window opened or closed during a lap, enabling Mercedes to fine-tune camber settings, pressure levels, and aero balance to mitigate overheating. His detailed input helped the team transition to setups that preserved tire life without sacrificing outright pace.

Fine-Tuning Novel Concepts and Systems (2020–2021)

The 2020 season saw Mercedes unveil the Dual Axis Steering (DAS) system. While conceptually it originated from the engineers, Hamilton's role in evaluating and optimizing its usage was openly acknowledged by the team. In interviews with Sky Sports F1 and the official Formula One website, Mercedes personnel praised how Hamilton explained where DAS benefited corner entry stability and tire warm-up. His clarity helped the engineers understand where and when the system provided a competitive edge. That year, Hamilton's concise guidance ensured that the innovative device delivered tangible performance gains, particularly in managing front tire temperatures on long straights.

In 2021, with Red Bull mounting a formidable challenge, the difference often came down to setup refinements. Hamilton worked closely with his race engineers, offering precise assessments of balance shifts under different fuel loads and wind conditions. Public statements from Shovlin and Wolff after qualifying sessions frequently referenced Hamilton's role in pinpointing the ideal wing configurations and suspension settings to counter evolving track conditions. Although the season ended contentiously, his feedback throughout allowed Mercedes to keep pace with their fiercest rival.

Influencing Fundamental Concepts (2022–2023)
The regulatory reset in 2022 introduced ground-effect floors and triggered severe porpoising for many teams, including Mercedes. Hamilton's post-race interviews

and radio communications, often quoted in official race reports and technical analyses, conveyed how the car's bouncing affected corner entry, braking stability, and driver comfort. Toto Wolff and Technical Director Mike Elliott confirmed in multiple media briefings that Hamilton's detailed descriptions of the phenomenon helped them identify which aerodynamic components were most responsible. His unwavering input guided the gradual evolution of the W13 and the conceptual shift that influenced the W14's development for 2023.

By 2023, as Mercedes introduced revised sidepod designs and other aerodynamic updates, the team reiterated Hamilton's importance in setting development priorities. In mid-season interviews, Wolff noted that Hamilton's comments on the car's traction out of slow corners and feedback on tire degradation patterns under the changed bodywork layouts helped the team refine their upgrade packages, inching them closer to the front despite ongoing challenges.

A Technical Voice in the Garage
From the moment Hamilton joined Mercedes, key figures within the team—Lowe, Allison, Shovlin, and Wolff—consistently highlighted his ability to translate on-track sensations into actionable engineering direction. Unlike some drivers who focus solely on lap time, Hamilton combined racecraft with a sophisticated

understanding of car dynamics. His clear, consistent feedback loops allowed Mercedes to respond quickly to regulation changes, tailor setups to new tire constructions, and implement novel systems effectively.

Across a decade of shifting rules and relentless competition, Hamilton did more than just adapt his driving style. He became an integral part of Mercedes' development cycle. His input steered everything from brake pedal feel in 2014 to aerodynamic concepts in 2023. In doing so, Lewis Hamilton not only enhanced his own performance but also actively shaped the technical trajectory of a team that defined an era of Formula One racing.

Chapter 20: Mentorship and Influence Within the Team

Beyond his role as a record-breaking driver and technical contributor, Lewis Hamilton exerted a subtler yet no less significant influence on the human fabric of the Mercedes AMG Petronas Formula One Team. Over the years, official interviews, team statements, and media reports documented moments in which Hamilton offered guidance to junior talents in the Mercedes driver program, encouraged reserve drivers stepping into new roles, and bolstered the morale of mechanics and engineers through personal acknowledgment and support.

Supporting Emerging Talents

One of the most notable examples of Hamilton's supportive stance involved the team's test and reserve drivers, many of whom would later graduate to full-time racing seats in Formula One. Esteban Ocon, who served as a Mercedes reserve driver in 2019, publicly acknowledged that observing Hamilton's approach and having the opportunity for informal advice helped him understand the depth of preparation required to compete at the front. While Ocon did not race for Mercedes at that time, he benefited from trackside interactions, debrief attendance, and the chance to study Hamilton's data. In interviews with Formula One's official media, Ocon referred to Hamilton as a benchmark whose work ethic and detail-oriented approach provided a template for young drivers aspiring to reach the sport's highest levels.

George Russell's early involvement with Mercedes—initially as part of their young driver program before securing a race seat with Williams—also illustrates Hamilton's mentorship presence. Russell, who joined Mercedes full-time in 2022, mentioned in multiple media interviews that he found Hamilton's willingness to share insights and perspectives invaluable, even prior to becoming teammates. Russell noted that Hamilton's candid explanations of car characteristics and his calm, analytical method of dissecting a race weekend made a lasting impression. According to Russell's statements quoted by Sky Sports and the official Formula One website, Hamilton offered advice on adapting to top-level racing demands, dealing with the intense scrutiny of the media, and maintaining composure under pressure.

Inspiring Morale and Unity Within the Garage
Mechanics, engineers, and other team members have often spoken about Hamilton's positive impact on internal morale. Toto Wolff, in interviews released through Mercedes' official channels, praised Hamilton's personal engagement with the crew. Wolff noted that

Hamilton regularly visited the factory at Brackley to meet with engineers and aerodynamicists face-to-face, asking questions and expressing genuine appreciation for their work. In behind-the-scenes Mercedes videos and official team interviews, technical staff have recalled Hamilton's habit of personally thanking them for upgrades or setup changes that improved the car's performance. Such gestures, team members said, created a constructive atmosphere where every contribution felt valued.

Senior engineers, including Trackside Engineering Director Andrew Shovlin, have remarked in post-race briefings that Hamilton's calm demeanour and encouraging words after difficult weekends helped steady the team. When facing setbacks—be it a problematic upgrade, unexpected reliability issues, or a challenging tire compound—Hamilton's ability to remain composed and offer measured suggestions rather than criticism helped maintain a positive working environment. His approach signaled that problems were solvable puzzles rather than failures, and this perspective kept morale high even during challenging phases of development.

Leading by Example

In numerous press conferences and documentary-style interviews, Hamilton emphasized that leadership within a team is not just about directives or grand speeches—it's about leading by example. His exacting standards in fitness, meticulous preparation for each race, and willingness to delve deep into data reinforced a cultural norm that hard work and attention to detail could yield results. Juniors and newcomers to the team—be they in the garage or in the simulator—had a role model who embodied professionalism. Official Mercedes communications often highlighted Hamilton's involvement in simulator sessions, where his presence and feedback helped less experienced drivers understand how to translate simulator runs into actionable insights on race weekends.

Beyond the technical and performance aspect, Hamilton's consistent recognition of the team's diverse efforts—be it aerodynamic refinements, efficient pit stops, or incremental reliability gains—fostered a collective pride. Wolff and other senior figures have commented publicly that Hamilton's admiration for the unseen hours invested by factory staff and traveling crew members made these individuals feel integral to the broader mission, motivating them to push their capabilities further.

An Enduring Legacy of Guidance
While Hamilton's influence on car development and race craft is often measured in terms of lap times and championship points, the mentorship and support he

extended to those around him also formed part of his lasting legacy at Mercedes. The official records, from press releases to team interviews, provide a clear narrative: through quiet counsel, empathetic feedback, and a willingness to share his wealth of experience, Hamilton helped shape not only the competitive fortunes of the team, but also the personal and professional growth of the people within it.

In this sense, Hamilton's time at Mercedes was about more than winning titles—it was also about nurturing the next generation of talent, strengthening the bonds within the garage, and inspiring individuals to reach new heights, both on and off the track.

Chapter 21: Hamilton's Evolving Brand and Off-Track Engagement

Throughout his years at Mercedes, Lewis Hamilton's public persona grew increasingly multifaceted and influential. Once regarded primarily as a supremely talented driver, he gradually embraced broader roles in philanthropy, fashion, and social advocacy. Official statements from the team and recognized media commentary confirm that these endeavors did not exist in isolation—they helped shape his personal brand and, by extension, the image of Mercedes as a forward-thinking and socially conscious organization.

Philanthropy and Advocacy for Diversity

One of the most documented expansions of Hamilton's off-track focus occurred around issues of diversity and inclusion within motorsport. In 2020, Hamilton established The Hamilton Commission, a research initiative partnered with the Royal Academy of Engineering, aiming to identify and address barriers to the participation of Black people in UK motorsport. Publicly announced and widely covered by prominent outlets like BBC Sport and The Guardian, the Commission's work ultimately produced a report with recommendations for educational and industry-level reforms.

Mercedes, for its part, issued official statements applauding Hamilton's efforts. Toto Wolff frequently spoke in interviews about how Hamilton's push for greater inclusion resonated with the team's own commitments. In 2020, Mercedes adopted a black livery as a symbol of solidarity against racial injustice—an action that Wolff and the team credited, at least in part, to Hamilton's leadership and advocacy. Media coverage from sources such as Sky Sports and CNN highlighted how Hamilton's stance on social issues elevated him beyond the realm of a champion driver and into an influential public figure.

In addition, Hamilton made contributions to global causes through his Mission 44 foundation, launched in 2021 with the stated aim of supporting, empowering, and uplifting young people from underrepresented backgrounds. Official press releases and interviews with Hamilton indicated that Mission 44 aligned with Mercedes' broader diversity goals, further entwining the driver's philanthropic identity with the team's evolving narrative.

Fashion, Lifestyle, and Environmental Stances
Hamilton's personal brand also extended into fashion

and lifestyle. Starting in 2018, Hamilton formed a high-profile partnership with Tommy Hilfiger, creating lines of streetwear-inspired clothing and occasionally appearing in brand campaigns. Media commentary in publications like Vogue and GQ noted that this collaboration showcased Hamilton as a style influencer—a driver who brought a fresh cultural dimension to a sport traditionally associated with conservative image-building.

His environmental advocacy gained traction as he publicly embraced a plant-based diet and spoke about sustainability issues. Interviews recorded by Formula One's official media channels and coverage in outlets like The New York Times documented Hamilton's emphasis on reducing his carbon footprint, cutting back on single-use plastics, and encouraging eco-friendly behaviors. While Mercedes did not explicitly recalibrate its entire image around environmental activism, team representatives acknowledged Hamilton's stance, framing it as reflective of the progressive values that a modern Formula One team could embody.

Aligning with Mercedes' Modern Identity
Official Mercedes communications during Hamilton's era frequently underscored the synergy between driver

and team identities. Mercedes ran campaigns that highlighted Hamilton's achievements beyond the track, sharing stories of his charitable work and social activism on their website and social media. In interviews, Wolff noted that Hamilton's personal growth aligned well with Mercedes' own evolution from a purely performance-driven constructor to an organization keen to promote social responsibility, innovation, and a connection to global issues.

As a result, Hamilton's off-track engagement created a positive feedback loop. His growing international following—spurred by his fashion ventures, music interests, public support for Black Lives Matter, and statements on equality—drew new audiences to the sport and the team. Official F1 fan surveys and reported social media engagement increases during Hamilton's peak years indicated that a portion of newly interested fans resonated with Hamilton's persona as more than "just a driver." The Mercedes brand, in turn, benefited from this broadened fan base, their image bolstered by the narrative of a champion who cared about representation, fairness, and the world beyond racing.

Media and Public Perception
Reputable motorsport journalists, as well as lifestyle and cultural commentators, increasingly referenced Hamilton as a global icon. Publications like The Times

and The Guardian analyzed how his outspoken support for racial justice and collaboration with high-fashion brands helped shatter the narrow stereotypes of what a Formula One driver could represent. No longer was he merely a winning figure on the track; Hamilton had become a multi-dimensional brand that championed social good, style, and personal authenticity.

Through these documented endeavors—philanthropy, advocacy, fashion partnerships, environmental awareness—Hamilton's personal brand came to reflect attributes that appealed to younger, more diverse audiences. The team's public backing of his initiatives validated these efforts and enhanced Mercedes' reputation as a dynamic force open to positive change.

A Legacy Beyond Lap Times
In retrospect, Hamilton's off-track activities show how a champion can leverage his platform to shape public discourse and inspire both fans and colleagues. Official team materials, interviews with Wolff, media commentary from respected journalists, and Hamilton's own public statements confirm that his brand trajectory contributed to redefining what it meant to be a Formula One superstar in the 21st century.

For Mercedes, this chapter of Hamilton's journey reinforced their identity as a team comfortable engaging in dialogues that transcended sport. The partnership thus extended beyond the garage and into the cultural domain, leaving a legacy of a driver and a team whose influence reached far beyond the checkered flag.

Chapter 22: Circuit Mastery and Signature Performances

While Lewis Hamilton's success at Mercedes spanned continents and calendars, certain circuits emerged as stages on which his talent shone with unmistakable brilliance. Whether it was his ability to manage tires in varying conditions, extract a crucial tenth in qualifying, or remain unflappable under pressure, these signature performances formed a tapestry of racecraft that defined his era at the team. Official statistics, race reports, and media coverage consistently highlighted how these standout showings helped cement Hamilton's reputation as one of Formula One's most versatile and formidable competitors.

Silverstone – The Home Advantage

If there was a track that became synonymous with Hamilton's Mercedes journey, it was Silverstone. Between 2014 and 2020, he amassed multiple British Grand Prix victories on home soil. In fact, he broke the record for most wins at a driver's home race, repeatedly displaying a near-supernatural ability to find grip and pace around the fast, flowing corners of the Northamptonshire circuit. During these years, he claimed poles and often secured fastest laps, illustrating a well-documented synergy with the high-speed demands of the track.

Journalists and team personnel alike commented that Hamilton's intimate familiarity with Silverstone's layout allowed him to adapt instantly to evolving conditions— be it sudden rain showers or gusty winds. The data-backed consistency in the corners of Maggotts, Becketts, and Chapel confirmed that his racing lines, honed from countless laps, gave Mercedes a strategic edge. After one particularly dominant win at Silverstone, Toto Wolff remarked in post-race interviews how Hamilton's confidence and precise feedback played into tailored aerodynamic and tire-pressure adjustments that left rivals trailing.

Montréal – A Historic Hunting Ground

Another circuit where Hamilton's name became etched in the record books was the Circuit Gilles Villeneuve in Montréal. Although his first Canadian Grand Prix victory came at McLaren in 2007, he continued his dominance there with Mercedes, notching multiple wins throughout the hybrid era. Trackside data and lap-time analysis from these weekends showed Hamilton's knack for braking stability into the final chicane and deft management of the tricky Wall of Champions. He regularly secured pole positions and converted them into race victories, often by controlling tire degradation and making decisive strategy calls.

In official Formula One post-race summaries, engineers highlighted how Hamilton's input on brake balance and gear ratios helped the team find an optimal setup for Montréal's stop-start nature. Such refinements paid off on Sundays, resulting in commanding drives that drew praise from commentators and peers. With each Canadian triumph, he reinforced a narrative of being especially adept at circuits requiring aggressive braking and a fine-tuned sense of rhythm.

COTA and the US Connection

The Circuit of the Americas (COTA) in Austin, Texas, also became synonymous with Hamilton's success at Mercedes. He routinely locked in wins and podiums there, making the US Grand Prix a venue where he could stretch his legs. The track's mixture of high-speed sweeps and slow hairpins seemed to play into Hamilton's strengths, allowing him to balance aggressive initial stints with careful tire conservation.

Official Mercedes debriefs often noted that Hamilton's clarity on where the car struggled helped the team tweak suspension settings and aerodynamic balance, ensuring maximal stability through COTA's undulating turns. Sky Sports F1 and other broadcasters repeatedly pointed out that the US crowd's enthusiastic support only further galvanized Hamilton, and his performance data—showcasing minimal lap-time drop-offs and intelligent energy deployment—underscored his mastery.

Hungry for More: The Hungaroring

Hungary's twisty Hungaroring, long considered a driver's circuit, stood as another arena where Hamilton thrived. Claiming multiple victories with Mercedes, he demonstrated an uncanny ability to excel at a track often compared to a go-kart circuit. The series of medium-speed corners demanded smooth inputs and patience, qualities that Hamilton leveraged to set up overtakes when opportunities arose. Official timing comparisons showed that Hamilton's sector times were consistently at or near the top in qualifying and that his race pace remained stable even as tires aged.

Team insiders and journalists alike highlighted Hamilton's methodical approach to the Hungaroring's challenges. In post-race interviews, his race engineers acknowledged that his detailed reporting on handling quirks—such as subtle understeer in mid-corner—led to incremental but meaningful setup improvements that directly translated into lap-time gains.

Defining a Legacy Through Dominance

These circuit-specific triumphs were not just isolated victories. Each contributed to Hamilton's evolving narrative as a driver capable of adapting to any environment. Official race reports, FIA statistics, and statements from both Mercedes and external analysts confirm that tracks as varied as Silverstone, Montréal, COTA, and the Hungaroring demanded distinct skills—

high-speed stability, heavy braking, intricate traction control, and deft tire management.

Hamilton's repeated success at such a diverse array of circuits bolstered his standing in the sport's pantheon. The data-driven evidence, from pole position counts to fastest laps, painted a picture of a driver who could dominate at home, thrive abroad, and excel in both classic and modern venues. By regularly overcoming unique challenges at specific tracks, Hamilton enhanced his legacy and gave credence to the view that he was not just a beneficiary of a strong car, but a masterful racer who raised the ceiling of what could be achieved when man and machine worked in perfect harmony.

Chapter 23: Adapting to Psychological and Competitive Pressures

From last-lap title deciders to the lingering disappointment of mechanical failures and mid-season performance dips, Lewis Hamilton's time at Mercedes was marked by numerous high-pressure moments. While his strategic acumen and car development input have been well-documented, Hamilton's evolution as a competitor also involved cultivating mental resilience and coping strategies. Drawing from interviews, team statements, and recognized media commentary, this chapter focuses on how Hamilton's psychological approach evolved over the years, illustrating the mindset shifts that enabled him to navigate the sport's intensities and uncertainties.

Overcoming Adversity in Title Showdowns
Hamilton faced some of the fiercest championship battles of the modern era. The 2016 title fight against Nico Rosberg, for instance, tested his emotional fortitude. After mechanical failures—most notably an engine blowout in Malaysia—undermined his campaign, Hamilton appeared outwardly composed in post-race interviews. He acknowledged frustration yet emphasized focusing on what he could control rather

than what he could not. In official press conferences, Hamilton spoke of channeling setbacks into motivation. Team personnel, including then-non-executive chairman Niki Lauda, publicly commended Hamilton's ability to rebound from disappointment, noting how he returned stronger and more determined to refine his driving, even under the heaviest emotional burdens.

Such instances revealed a pattern: while in earlier years Hamilton might have shown more visible exasperation over misfortunes, by the late 2010s he consistently portrayed calm acceptance in the face of bad luck. Speaking to F1's official media and outlets like the BBC, he described how personal growth and life experience taught him to confront setbacks with measured perspective. This shift suggested that, with time, he learned to harness adversity as fuel rather than letting it derail his performance.

Finding Stability Amid Mid-Season Slumps
Not every challenge revolved around titles. Occasionally, Mercedes endured performance dips—whether due to tire struggles, aerodynamic miscalculations, or unexpected rival surges. During those periods, Hamilton's mental resilience was often cited by senior team figures in interviews with motorsport publications. Toto Wolff noted that

Hamilton's calm demeanor when results faltered helped stabilize the garage atmosphere. Instead of lashing out, Hamilton focused on incremental improvements. He publicly stated that when chasing form, he leaned on established routines: careful data review, open communication with engineers, and maintaining strict fitness and rest regimens to keep his mind sharp.

Sports psychology professionals associated with the team rarely granted in-depth interviews, but in brief comments, Mercedes staff acknowledged that a key to Hamilton's longevity at the top lay in his capacity to manage stress internally. Although specifics remained private, Hamilton himself alluded to methods he employed—such as mental visualization techniques and deliberate breaks from social media noise—during interviews with Formula One's official channel. He indicated that these techniques helped maintain balance, ensuring that even during challenging weekends, he retained the clarity needed to provide constructive feedback and commit to long-term improvement.

Evolving Coping Mechanisms and Personal Growth
Over time, Hamilton became more open in discussing mental well-being. In interviews with outlets like Sky Sports and CNN, he talked about personal growth—

reading motivational literature, seeking inspiration outside racing, and drawing on his interests in music and travel to replenish his mental energy. Angela Cullen, his long-serving physiotherapist and performance coach, often mentioned in team Q&As, played a role in helping Hamilton maintain a consistent psychological framework. While never delving into clinical detail, media coverage noted how Hamilton credited his close-knit support circle for helping him stay centered amid intense scrutiny and expectation.

Public commentary also highlighted how Hamilton embraced positive thinking and reframed setbacks as lessons. In 2018 and 2019, when battling Ferrari and Sebastian Vettel, Hamilton frequently discussed how he learned not to be rattled by early points deficits. By calmly reiterating his trust in the team's process and acknowledging the season's length, he neutralized psychological pressure. Instead of fixating on immediate losses, he spoke about the "bigger picture," a term he used repeatedly in official race debriefs and interviews. This reframing technique, widely reported by motorsport press, allowed him to maintain composure under the brightest spotlights.

Influence on Perception and Legacy
Media and commentators recognized that Hamilton's evolution in handling pressure contributed to shaping

the narrative around him as a complete athlete. Early in his career, questions sometimes arose about his emotional control under stress. By the height of his Mercedes tenure, journalists and analysts in publications like Autosport and Motorsport.com described him as a driver who matured emotionally. They linked his remarkable consistency, both in good times and bad, to the mental tools he had honed.

This growth in psychological resilience, supported by evidence from team statements and Hamilton's own words, reinforced his standing as not just a natural talent, but a competitor who continuously refined his mental approach. The calm he displayed after engine failures, the steadiness amid title showdowns, and the proactive steps taken to ward off negativity all contributed to a legacy in which mental strength paralleled physical skill.

A Mindset Shaped Over a Decade

In sum, Hamilton's trajectory under Mercedes was as much about mastering psychological pressures as it was about achieving sporting greatness. Recorded interviews, official press conference notes, and team-approved insights confirm that he learned to remain composed in adversity, find stability during slumps, and turn defeats into sources of renewed motivation. By forging these mental habits and coping mechanisms over a decade at the pinnacle of motorsport, Hamilton showcased that his success stemmed not only from engineering mastery and race craft, but also from a profoundly resilient and ever-evolving mindset.

Chapter 24: A Farewell and a New Horizon

As the curtain fell on the 2024 season, Lewis Hamilton's decade-long chapter with Mercedes drew to a close. Coming off a year of incremental improvements in 2023, hopes had been modest yet persistent that the team might recapture some of its old form. But the new regulations, evolving competition, and a car that never quite found the sweet spot meant that Hamilton's final campaign in silver and teal was, by the team's historical standards, a struggle. Race after race, the W15 lacked the precision handling and outright pace to challenge for podiums regularly, and Hamilton found himself working harder than ever for every point. The season ended with him seventh in the Drivers' Championship—his lowest ranking since joining Mercedes.

While Hamilton and the team had certainly endured tough patches before—2016's championship loss, the porpoising debacle of 2022, the winless season that followed—2024 felt different. The competition around them had intensified further, with Red Bull maintaining their place at the top and other teams surging. It became clear early on that no simple solution would vault Mercedes back to title contention before the year was out. The aerodynamic upgrades helped in some respects, the power unit was solid, and the team's trackside operations remained professional and precise. Still, the combination never gelled into a winning formula. Hamilton, always a fighter, continued to extract what he could, securing respectable finishes and consistent points, but never reaching the top step of the podium that had once seemed almost synonymous with his name.

As the season wound down, the knowledge that 2024 would be Hamilton's last with Mercedes settled into the atmosphere of the garage. Rumors had swirled for months about his future, and by mid-season, the announcement that he would move to Ferrari in 2025 removed any lingering doubt. This was no hasty decision; it was the conclusion of careful thought, private negotiations, and an understanding on both sides that the time had come for a fresh start. Mercedes, having reaped the rewards of Hamilton's prime years—countless wins, poles, and six Drivers' Championships with the team—expressed gratitude rather than resentment. The relationship, forged in triumph and tested by adversity, ended with mutual respect.

The final race weekend of 2024 felt charged with emotion. While Hamilton had bid farewell to previous teams before, this departure carried the weight of a historic era coming to an end. From the fabled 2014 breakthrough to the record-breaking feats that followed, Hamilton and Mercedes had set standards against which all future success would be measured. In the paddock, rival team principals, drivers, and engineers paid tribute. Journalists, who had covered his rise from a prodigious talent at McLaren to an all-time great at Mercedes, noted that this closing moment marked the end of Formula One's most dominant partnership in the 21st century.

In the garage, mechanics and engineers—some of whom had worked with Hamilton since his arrival—lined up to shake his hand. Many had been there through the highs of championship celebrations and the lows of mechanical heartbreak. They had overcome regulation shifts together, braved intense title battles, and weathered the disappointment of near-misses. The unity and trust built over more than a decade, though now shifting course, would not simply vanish.

On the grid before the last race, Hamilton stood beside his car, the emblematic three-pointed star still gleaming under the track lights. He took a moment to reflect: on his record-smashing statistics, the technical triumphs, the strategic masterstrokes, and the personal growth he experienced in this environment. Emotions were visible—an acknowledgment that even legends of the sport feel the weight of endings. He climbed into the cockpit knowing this would be his final performance for the team that had defined his career's pinnacle. The race itself was a microcosm of 2024: hard-fought, carefully managed, but ultimately short of the podium. When the checkered flag fell, the journey that began in 2013 concluded not with a triumphant roar, but with a quieter note of respect and closure.

After the race, Hamilton embraced Toto Wolff and the senior figures in the team. He praised the mechanics for their tireless efforts and thanked the engineers and strategists who had worked so long and hard to return the squad to the front. Although results in 2024 had been elusive, no one doubted the sincerity of his gratitude or the magnitude of what they had achieved together over the years. In the press pen, media commentary struck a reflective tone. Some focused on the poetic symmetry of a great champion turning a page; others looked ahead, curious to see how Hamilton would fare in red, and how Mercedes would chart a new course without him.

As the lights dimmed and transporters began to pack away equipment, the legacy was set. Lewis Hamilton had transformed Mercedes into a juggernaut, and Mercedes had provided him the platform to become one of the most decorated drivers in history. Their time together had reshaped records, redefined standards, and influenced global fan engagement. Safety, strategy, engineering prowess, and sporting values had all evolved during this era. Now, as Hamilton looked ahead to Ferrari and Mercedes prepared for life after their star driver, the story ended not with bitterness, but with the quiet dignity of two legendary entities parting ways, each forever enriched by the other.

Chapter 25: A Gallery of Iconic Moments

In this final chapter, we present a curated photo gallery hinting at images that capture key moments from Lewis Hamilton's era at Mercedes. While this is a text-based format, imagine these as carefully selected photographs that tell the story of his journey, each requiring only a few words to conjure the memory for those who know the history.

1. **2013 Debut in Silver:**

2. **The First Win for Mercedes (Hungary 2013):**

3. **Championship Glory (Abu Dhabi 2014):**

4. **Dominant Partnership (2015 Season Celebration):**

5. **The Duel in the Desert (Bahrain 2014 & 2016):**

6. **Porpoising and Persistence (2022):**

7. **Breaking Schumacher's Win Record (Portugal 2020):**

8. **Triple-Header Triumphs (The W07 Era in 2016):**

9. **The Halo and Changing Safety Standards (2018):**

10. Black Lives Matter Livery (2020):

11. 100 Poles and 100 Wins (2021 Milestones):

12. The Narrow Defeat (Abu Dhabi 2021):

13. Steady Hands in Hard Times:

14. The Final Farewell (2024 Finale):

15. **Next Chapter Beckons (2024 Post-Race Scene):**

Each photograph represents a fragment of Hamilton's Mercedes story. From debut to departure, from record-breaking feats to final farewells, these images would illustrate a legacy defined by skill, resilience, emotion, and the unwavering pursuit of excellence.

About the Author

Etienne Psaila, an accomplished author with over two decades of experience, has mastered the art of weaving words across various genres. His journey in the literary world has been marked by a diverse array of publications, demonstrating not only his versatility but also his deep understanding of different thematic landscapes. However, it's in the realm of automotive literature that Etienne truly combines his passions, seamlessly blending his enthusiasm for cars with his innate storytelling abilities.

Specializing in automotive and motorcycle books, Etienne brings to life the world of automobiles through his eloquent prose and an array of stunning, high-quality color photographs. His works are a tribute to the industry, capturing its evolution, technological advancements, and the sheer beauty of vehicles in a manner that is both informative and visually captivating.

A proud alumnus of the University of Malta, Etienne's academic background lays a solid foundation for his meticulous research and factual accuracy. His education has not only enriched his writing but has also fueled his career as a dedicated teacher. In the classroom, just as in his writing, Etienne strives to inspire, inform, and ignite a passion for learning.

As a teacher, Etienne harnesses his experience in writing to engage and educate, bringing the same level of dedication and excellence to his students as he does to his readers. His dual role as an educator and author makes him uniquely positioned to understand and convey complex concepts with clarity and ease, whether in the classroom or through the pages of his books.

Through his literary works, Etienne Psaila continues to leave an indelible mark on the world of automotive literature, captivating car enthusiasts and readers alike with his insightful perspectives and compelling narratives.

Visit **www.etiennepsaila.com** for more.

www.ingramcontent.com/pod-product-compliance
Ingram Content Group UK Ltd.
Pitfield, Milton Keynes, MK11 3LW, UK
UKHW050413241224
452714UK00003B/12